Organic Pest Management

Potentials & Applications

Organic Pest Management

Potentials & Applications

Editor

Dr. P. NARAYANASAMY

Dean, Faculty of Agriculture & Prof. & Head,
Department of Entomology, Annamalai University
Annamalai Nagar - 608 002. Tamil Nadu

SSPH

SATISH SERIAL PUBLISHING HOUSE

403, Express Tower, Commercial Complex, Azadpur, Delhi-110033 (India)
Phone : 011-27672852, Fax : 91-11-27672046
E-mail :info@satishserial.com, hkjain1975@yahoo.com
Website : www.satishserial.com

Published by :

SATISH SERIAL PUBLISHING HOUSE
403, Express Tower, Commercial Complex, Azadpur, Delhi-110033 (INDIA)
Phone : 011-27672852 Fax : 91-11-27672046
E-mail : info@satishserial.com, hkjain1975@yahoo.com

Reprint 2018

₹ 2950.00

ISBN 81-89304-75-5

Typeset at: **Laxmi Art Creations** # 98 11 48 23 28
Printed at: Shri Ganpati Enterprise, Delhi

डॉ. ओ. एम. बम्बावाले
निदेशक

Dr. O.M. Bambawale
Director

भा.कृ.अनु.प.
ICAR

NCIPM

राष्ट्रीय समेकित नाशीजीव प्रबन्धन केन्द्र,
(भारतीय कृषि अनुसंधान परिषद)
लाल बहादुर शास्त्री भवन, पूसा परिसर, नई दिल्ली-110012

NATIONAL CENTRE FOR INTEGRATED PEST MANAGEMENT
(Indian Council of Agricultural Research)
L.B.S. BUILDING, PUSA CAMPUS, NEW DELHI-110012

Foreword

Agriculture in the country has underwent extensive changes in the crop protection sector through varieties of eco-friendly tactics and technologies towards ushering in an era of organic pest control. Considering this concept, the Department of Entomology, Faculty of Agriculture, Annamalai University, Tamil Nadu organized optly a maiden training programme for Assistant Professors, subject matter specialists and Junior Scientists on a topic "Organic Pest Control" during January 21st - February 10th, 2008.

I feel happy to observe that lectures were organized on various important topics viz., Biopesticides, pheromones and traps, bio-control agents, flyash etc., and such papers have been compiled together in the present book form. This is an excellent piece of material to come up at a right time in the country wherein we have started practicing organic methods of pest management. I would appreciate and congratulate my friend - scientist, Dr.P.Narayanasamy, Dean, Faculty of Agriculture and Professor & Head, Department of Entomology, Annamalai University for the efforts to organize the training event and for getting the book to reality.

I foresee that the book will be a very good reference and text material for those who practice biological pest control methods in the field, teachers and students of Entomology, Plant Pathology and Nematology.

Dr. O.M.Bambawale

Preface

In an attempt to promote eco-friendly ways of pest control in agriculture, the University Grants Commission, New Delhi approved our proposal of a 21-day Refresher course on a theme, "Organic Pest Control", which was held at the Department of Entomology, Annamalai University during January 21st – February 10, 2008. This had trainees from various institutions from Mizoram, Gujarat, Kerala, Maharastra, Orissa besides Tamil Nadu. Experts of various fields like organic pest control scenario in the country, microbial and herbal pesticides, lures and traps, non-toxic substances like flyash, host plant resistance etc., gave lectures and the papers have been compiled here. My profound thanks are due to the UGC, New Delhi for funding the programme without which this book would not have seen the light of the day.

I am much thankful to our respected Founder – Pro-Chancellor, Dr.M.A.M.Ramaswami and honourable Vice-Chancellor, Dr. M. Ramanathan for the permission and support rendered to conduct the training programme.

My deep regards are due to our esteemed Registrar of Annamalai University, Dr.M.Rathinasabapathy for providing all logistics and support.

I am much privileged to thank Dr.T.P.Rajendran, Asst. Director General, ICAR, New Delhi for participating in the Inaugural function and for delivering key lecture.

I am delighted to thank Dr.O.M.Bambawale, Director, National Centre for Integrated Pest Management (NCIPM), ICAR, New Delhi for delivering Valedictory Address and a special lecture. Also I cherish with happiness to thank him for his Foreword to this book.

Grateful regards are due to Dr.O.P.Sharma, Deputy Director, NCIPM, ICAR, Dr.P.S.Vimaladevi, Directorate of Oil Seeds Research, Rajendranagar, Dr.D.Rajagopal, Director of Research (Retd.), UAS, Bangalore, Dr.S.Sithanantham, Sun Agro Bio Systems, Chennai, Dr.Chandish Ballal, National Bureau of Agriculturally Important Insects (NBAII), Bangalore, Dr.Opender Koul, Insect Biopesticide Research Centre, Jallandhar and Dr.V.Nandhagopal, Central Rice Research Institute, Cuttack for contributing to this volume.

Let me thank all my fellow colleagues in the Faculties of Agriculture and Science in the Annamalai University, Tamil Nadu Agricultural University and a score of Biopesticide establishments for the support.

My heartfelt appreciations and regards are due to M/s Satish Serial Publishing House, New Delhi for focusing the theme in the form of a book with an excellent get-up.

DR.P.NARAYANASAMY
DEAN, PROF. & HEAD
FACULTY OF AGRICULTURE
DEPARTMENT OF ENTOMOLOGY
ANNAMALAI UNIVERSITY

Contents

Foreword *v*
Preface *vii*

I. Organic Pest Control–Scenario in India

1. Status and Scope of Organic Pest Control in Indian Agriculture 1
Rajendran, T.P.

2. Scope of Organic Pest Control Options in Integrated Crop Pest Management 11
Sharma, O.P., D.K. Garg & O.M. Bambawale

3. Scope of Organic Pest Control in Oilseeds Crops 25
Vimala Devi, P.S.

4. Eco-friendly Management of Soil-Borne Pests 37
Rajagopal, D.

5. Organic Pest Management in Agricultural Ecosystems 47
Nadarajan, L.

6. Integrated Pest Management in Tea in South India: Problems and Prospects 65
Babu, A.

7. Pest Problems and Their Eco-friendly Management Techniques in Cardamam 75
Kumaresan, D.

8. Management of Coconut Pests under Organic Farming Conditions 89
Ramaraju, K.

9. Promoting IPM with Focus on Organic Pest Control ... 103
Hariprasad, Y.

II. Utilization of Biological Control Agents

10. Emerging Trends in Improving the Impact Potential of Biocontrol Agents: Case Study of *Trichogramma* 129
Sithanantham, S.

11. Biological Control of Pests in Horticultural Ecosystems 133
Krishnamoorthy, A.

12. Mass Production Protocols for Potential Biological Control Agents 147
Chandish Ballal & Sunil Joshi

13. Improved Mass Production Technologies of Major Biocontrol Agents for Efficient Pest Management 163
Vijayakumar, N.

14. Analysis of Natural Enemy Biodiversity in Rice and Cotton Ecosystems & Evaluation of IPM Components 177
Ananda Ganesa Raja, B.

III. Utilization of Microbials

15. Utilization of Microbials in Crop Protection 187
Bambawale, O.M.

16. Registration for Mass Production of Microbial Biopesticides 197
Vimala Devi, P.S.

17. Impact of Research and Development (R & D) in Enhancing the Field Potential of Baculoviruses 203
Narayanan, K. & G.P. Shetty

18. Mycoinsecticides : Towards Organic Pest Control 219
Narayanasamy, P.

IV. Botanical Insecticides

19. Potential of Botanical Biopesticides for Efficient Pest Management Strategies 227
Opender Koul

20. Secondary Plant Compounds in Insect Pest Management Practical Experiences and Prospects 239
Arivudainambi, S.

21. Utility of Plant Products in Organic Pest Control 249
Veeravel, R.

V. Pheromones and Traps

22. Use of Pheromones for Managing Pests of Field Crops 257
Nandagopal, V.

23. Traps for Management of Stored Product Insects 267
Mohan, S. & P. Pretheep Kumar

24. Use of Lures and Traps as Ecofriendly Techniques for Organic Pest Control: Case Study of Fruit Flies 275
Sithanantham, S.

25. Traps in Crop Pests Management 279
Kathirvelu, C.

VI. Host Plant Resistance

26. Host Plant Resistance as Complement to Organic Pest Management 297
Selvanarayanan, V.

VII. Flyash in Pest Control

27. Flyash : Potential Component in Organic Pest Control 305
Narayanasamy, P.

VIII. Phytonematodes

28. Potential of Biocontrol Agents in the Management of Nematode Pests of Crops 315
Jonathan, E.I., S. Prabhu, K.Devrajan & P. Senthilkumar

29. Scope of Organic Amendments for the Management of Plant Parasitic Nematodes 325
Jonathan, E.I., S. Prabhu, K. Devrajan & P. Senthilkumar

IX. Miscellaneous

30. Appropriate Plant Nutrition Technologies for Sustainable Pest Management 333
Ravichandran, M

31. Plant-Pest Interaction 345
Panneerselvam, R.

Index 351

SECTION - I

ORGANIC PEST CONTROL – SCENARIO IN INDIA

1

Status and Scope of Organic Pest Control in Indian Agriculture

RAJENDRAN, T.P.
Assistant Director General (Plant Protection), Indian Council of Agricultural Research, Krishi Bhavan, New Delhi-110 001.

1. Introduction

Organic farming has taken roots in modern global agriculture and India has also established its traditional faith and practices in this 'new' connotation of farming in last two decades (Kalisko and Kalisko, 1939). It is heartening to note that there are a number of states in the country that have boldly declared that they would pursue policies on organic farming with attendant facilitation to their farmers in both practising the process of farming and marketing. The preamble of organic farming has been aimed at conservation and optimised utilisation of all natural resources for a reasonable profitability under the guiding factors of sustainability of the farm. In order to keep a certain threshold of profit from the farms, all the farming practices have to be redesigned to undo the ill-effects that have crept in the current agricultural scenario while attempting to increase cotton production in the prevalent cropping systems. Profitability under this scheme of things then connotes the very sustenance of the parcel of soil that is the basic capital of the farmers.

2. Early Experimentation in Cotton (Rajendran and Venugopalan, 1999; Rajendran *et al.*, 2000)

Although it was a talking point in the eighties of the last century, organic farming became a wide spread approach in the late 90s. I commenced its experimentation in Central Institute for Cotton Research in 600 m^2 plots with a highly pest-susceptible, but chosen quality fibre-yielding cotton variety, LRA 5166 since 1991. One of the premise that emboldened me to take up this study was my concurrent observations in previous year of selective egg-laying pattern of American bollworm moths and lower susceptibility to sap sucking insects such as jassids in certain plots in 12 year old long-term manurial trials. Recurrent observations in these experiments proved that the intrinsic rate of multiplication of jassids could be brought down to 0.18 from 0.45 or at times 0.60 in the same variety with recommended fertiliser applications. The decade analysis of wide-spread cultivation under organic farming practices led to the seed cotton yield as given in Table 1. The stability in cotton yield is commendable in spite of the vagaries in rainfall as well as other in-built uncertainties of dry-land farming (Doroth Myers and Suer Stolon, 1999; Tarhalkar *et al.*, 1996; KVM, 2004).

Table 1 : Seed cotton productivity in rainfed organic farms in Maharashtra (mean of 5 districts)*

Year	Status	Seed cotton (kg/ha)
1	Conventional (chemical-based)	500
2	1st year of conversion	575
3	2nd year of conversion	625
4	3rd year of conversion	750
5	4th year Organic farms	875
6	8th year Organic farms	1000
10	10th year Organic farms	1250

*Data of 1992-95

3. Development of Organic Farming in the Country

With the existing knowledge base about the procedures of organic farming of various crops out of the conventional knowledge base as well as from the advances made due to various research developments in our research institutions. The initiatives taken by Bombay Burma

Trading Corporation (BBTC) to convert Singampatti group of estates in southern India to market organic tea internationally is the first of the few efforts for commercial organic agriculture. A number of organisations such as The Ecological Development Society in Puthuchery, Institute for Integrated Rural Development at Aurangabad, The Society for Equitable Voluntary Actions (SEVA) in West Bengal, The Indian Agency for Organic Agriculture (IAOA), Peekay Tree Crops Development Foundation (PTCDF) at Cochin undertake training of personnel towards organic farming. All India Federation of Organic Farmers promotes organic farming in the country (IFOAM-ASIA,1997; Thampan, 1995; Planning Commission, 2002, ICAR, 2006).

4. Modernised Agricultural Production Strategies

The thesis on which modern agriculture has been balancing its approaches for higher farm productivity over seasons of one year rests on a cogent relationship between input-output supply of farms. The equation that is built up on these two factors for attaining profitability is around two variables, i.e., soil and weather (season). The emphatically exerted influence of these two variables in crop productivity has been less focussed while calling for fulfilling high productivity through proportionate input-based agriculture. Indian farming has been led to the input-based crop production system without any sensitivity to weighing out the appropriateness or universality of inputs in every given situation in this vast country that has a continental status in terms of climate and soil variability (Rupela *et al.*, 2006; Pfeiller, 1973).

The basic struggle of farmers for one season's crop production could be traced to maintenance of ideal plant population per unit area and then making that to produce what he dreams to harvest after accounting for seasonal herbivory (including insects, diseases, nematodes and the host of biotic issues) as well as climatic turbulence. The net harvest (after keeping his stocks for the family for the year) is what is marketable and is thrown up to market turbulence of supply and demand-based price fluctuations. So, their real struggle of average rainfed farming is to sustain three predictable stresses of every season in terms of herbivory, climate and market. These become extremely unpredictable for the farmer when his economic power is so fragile (Kairon *et al.*, 1998a,b).

5. Farming Steps

Much of this discussion shall be on the dry-land farming background. This is because the irrigated land has got the biggest risk-avoidance in terms of water, being at the feet of the farmer and he can mould his requirements of farming, based on his ability to harness this important farming resource. The algorithm of crop production commences with the tilling of land for which he needs the blessing of pre-monsoon or monsoon rains. To work on the dry-land, vertisol soils with country plough that is the best bet even now (with most advanced agricultural engineering innovations of farm machinery), adequate moisture that has to be equally preserved, with deft balancing act of age-old lessons of conservation, so as to enable sowing operations and permit germination of the given crop species. Once sowing of DESIRABLE crop variety is decided, based on the given rainfall pattern and prudent prediction that the farmer has to make, based on the hoard of backed up experienced prediction and its potential success, about early maturing, mid or long season crop variety seeds. Another most crucial decision is also about the market acceptability of that variety as an agricultural commodity by the elite consumers who throw the stones of desires in the still waters of market forces, causing ripples of uncertainties to build up into waves of curse of fluctuating price for the farmer by the time he harvests the crop. Thus the proverbial risk-intensified dry land farming is prevailing realilty in this country as a major grain-bowl for humans, animals as well as for many hundreds in the food chain as well as value addition chain. How well we developed solutions to conquer this huge risk of a fragile farmer who has no capital-base to sustain short or medium term financial and economic shocks. And if he decides to wean out his children from this glorious endeavour that our country has, from *vedic* times, been professing to employ for national prosperity over several centuries, it is just because of the harsh economic compulsions and challenges of modernity that he has to face as an additional and the most forceful threat along with those from his land or all other village-affiliated assets (Tandon, 1994).

The clamour for judiciousness has been the best sermon in recent years, being floated to mitigate this economic fracture that the farmers sustained over decades. This is compounded by the immense shortage of labour at the most crucial farm operations has been pernicious due to negative farm growth and poor paying capacity of farmers to sustain the floating labour population that supplemented his own family labour component. This chunk of village labour force has become migrant labour that gets income that is more lucrative from the booming urbanisation.

6. Risks Associated with Package of Practices in Organic Farming

Dose based input servicing is what modern agriculture patronised with the view that inputs determine the output of the farm unit. Since the entire game of farming has several externalities, it is too difficult to seek any balancing act. However, certain adage-based theories that have come out of beaten track of rustic prudence shall be thumb-rules for keeping the break-even situation a lively hope for the farmer on a decade-analysis of his endeavour. This becomes significant when one could se several thousand 'green' products being marketed in the country in the name of safer, eco-friendly inputs for organic and other types of farming. One of the tenets of organic farming is the internalisation of resources including inputs that are essential to nurture soil, crops and protect them from pestilence and herbivory. The word 'prudence' has been specially alluded to emphasise on the avoidance of carbon-copying of the agro-techniques as universal realities to be applied in the given farm situation. The farmers have to be weaned out of the habit of market-driven input supply chain. The villagers shall have to come together to produce or acquire the various inputs that are required for their farms. Based on the social, attitude, cultural and community aspects there are definitely anticipated pragmatic problems in realising this ideal situation in which there shall be economy and insurance of quality for all the products that have to be utilised in their farms.

7. The Following Practices shall be Worth Consideration

1. Internalise all the farm inputs is the basic principle for sustaining pressure against overpowering tendency of dumping by input dealers. All farm resources shall be converted as farmers' capital to resist this process. To forge the resources, modern scientific practices of composting and recycling of farm residues may be introduced so as to remove the word 'WASTE' from farming lexicon. The best of techniques using fungi (*Trichoderma* Spp., *Plurotus* sp., etc., bacteria (*Pseudomonas, Aspergillus* etc. or earthworms (*Eudrillus* or *Eisenia* or the like) or any such promising techniques that could be adapted according to the size of the

farmers' pocket. It ultimately adds on to the farmers' capital for undertaking farming. Recycling of farm wastes should be integrated with water-harvesting techniques so that dry land farming could become self-sustaining.

2. The immense threat due to herbivory shall be mitigated through two-pronged approaches.

- Cultivate crops with graded nutrition that adequately sustain their metabolism and functions to provide a decent crop yield. Surges of nutritional status due to top dressing philosophy shall provide open invitation to herbivores that prowl around only to harness an opportunity to get a decent life and to generate a new healthy generation. If the nutritional requirement at different growing phases (phenology) for sustaining optimum metabolism of crop species is identified, accordingly the farmers could optimise the crop nutrition and economise on losses from both natural forces (leaching, volatilising etc.) as well as from the luxury consumption by plants that are then depredated by a host of organisms that are branded as pests. Do we have similar pest episodes in the wild and natural systems? So, then, why is it that agricultural lands become so special for these organisms who adapt well from the wild and get the opportunity to flourish well over a few generations in modernised farms?
- Could be take up suitable strategies to also ward off those challenging herbivore species by adopting to shrewd techniques that induce avoidance, antibiosis, deterrence or even kill of certain fragile life stages? After all, pest management is ultimately reduction of herbivory by both reducing their population and consequent reduction in damage to the crops in question. Saved crop harvest is the increased harvest that farmers are enabled. This principle has to be encashed by growers with extreme diligence and care.

3. Enabling the optimised production through minimised wastage of farm resources has to be the best solution that farmers strive for so as to reduce avoidable expenditure. Farm resources are costlier and difficult to replenish.

8. Community Farming or Producer Companies (Co-operatives)

Yet another major approach for improving the efficiency of crop production and protection of farms from noxious organisms is by collective and co-operative steps at village level utilizing the expertise available for this purpose. Synchronised and planned action would always be more beneficial to improve the effect of human interventions for ideal crop protection. Decision making regarding suitable and best steps for improving the efficiency of plant protection operations is possible when the entire geographical region, one village or a group of them, as the case may be, would initiate steps in this regard. If insect suppression in cotton has to be detached from constant and injudicious use of insecticides in the crops, much attention has to be paid to conserve natural enemy activity in the methods in a geographic area. A harmonious pest management programme can be developed only by the integration of biological, chemical and cultural suppression methods, in addition to co-operative efforts of farmers, extension and research staff of all institutions in the region.

9. *Jaivik Krishi* – The Theme for Twenty-first Century Farming Stability

The latest government policy on agriculture initiatives also addresses the organic means of cultivation. Recent policy planning saw the encouragement of concepts of producers companies with strong internal control systems that enable self-certification with suitable national and international accreditation. Participatory certification system has been in vogue in order to have label claim for each farm or farming groups. These could obviate strong cost erosion through commercial certification. Organic crop production offers better judiciousness and prudence in optimized resource channelisation. Further, in the current opportunistic marketing system, this would offer farmers rightful farm profitability. Cost effective crop production shall be the accent of the current decade of this millennium to thwart all extrinsic forces that impede profitability of average rainfed farmers. Irrigated farmers also have the lessons to be used in more magnified way since their capital intensity of farming make them more vulnerable to the shocks that are impinged upon season after season.

The urgent need of agriculture to be, on one hand, able to fulfil demands, quality and safety, in terms of food security and affordability and to be at the same time environmentally sound and socially acceptable has increased in the opinion of both, scientists and the general public. *Jaivik Agriculture* (JA) is product and market driven, based on quality to be safe with labelled products. Being still a fast growing food sector, it is difficult to speculate about the size of the market volume in the future. JA requires the use of on farm inputs, organic seed, crop-rotation and traceability through internal control systems (ICS).

The Production Companies (PCs) that are fortified with internal control systems as well as Participatory Guarantee System shall be able to carry forward the standardised procedures and protocols of organic production of candidate commodities. The marketing approach as well trade benefits shall be shared by the participating farmers of the designated PCs. The enabling process of domestic certification standards under this self-regulating scheme shall obviate high costs of certification and could offer the domestic market a label claim for the organic products.

Organic waste recycling and crop health management are the cornerstones of successful farming that could be practised in holistic farming. The prospects of promotion of holistic agriculture, as I would now connote this transition in Indian agriculture from intensive, chemical-based to organic, biodynamic or vedic farming, are in an evolving stage in the country. Many agencies including those in government are striving for perfecting policies and road maps during the tenth five-year plan of the country. Recent initiatives by the Department of Agriculture (Integrated Management Division), Ministry of Agriculture of Government of India and the Food and Agricultural Organization of the United Nations direct the tone of the country's future policy in this regard. Indian Council of Agriculture organised National Conference on Vedic Agriculture and triggered opportune approaches to widely patronise the holistic agriculture in our country to ensure sustainability of our farms. Farmers have to be organized into Producer Companies to prevent fragmentation of land as well as to sustain competitive and market-driven agriculture.

The suggested support system from government to promote the organic farming as an underlined sustainable farming practice may be given following thrusts.

- Provide financial and input support during transition stage of 2-3 years to farmers who want to convert to organic farming. Government can design subsidies and promotional schemes for the same. Subsidy is to be extended through women SHGs - subsidies on cows, bullock-carts, bio-gas units, vermicompost etc. The subsidy taken away from chemical fertilizers may be diverted to promotion of vermicomposting, recycling of farm wastes and other efforts to boost soil nutrition and stability. Market or premium may also be guaranteed during the shift from chemical fertilizer based agriculture to organic farming. Institutions like NABARD can help by extending long-term crop loans schemes for organic crops and marketing strategisation along with premier management institutions of the country.

- Allocate at least 30% of the central and state agricultural budget to organic farming. Govt.'s commitment to this area may be visible in its policies, decisions, schemes as well as budgetary allocations.
- Integrate natural resource management programmes (watershed, biodiviersity, tree plantation, forestry development, rural development) to promote organic farming.
- Non-government agencies & Self help groups may be richly involved for grass-root implementation, sensitize, train and build capacity of the Agricultural Extension staff, create local seed banks for production and multiplication and promotion of traditional varieties and for enabling recycling of farm wastes.
- Patronize crop health management as the major theme of agriculture instead of breaking it into ICPM, IPM and other sub-sets that confuse farmers and make him forget the centrality of farming.
- The emergent need is to dovetail all agricultural production practices that support better rhizosphere management and bring about stable crop production environment under farmers' field conditions. The best soil health leads to good plant health and the best metabolic activities in crop plants to attain genetic potential of crop yields.

Acknowledgement

I thank the Indian Council of Agricultural Research for the logistics and opportunities.

References

Carson, R. 1962. Silent Springs. Boston, M.A. Houghton Miffin 368p.

Dorothy Myers and Suer Stolon (eds.) 1999. Organic cotton from field to final product. Intermediate Technology Publication, 250 p.

IFOAM-ASIA, 1997. *Third IFOAM-ASIA Scientific Conference "Food Security in Harmony with Nature"*, 1-4 Dec., 1997, UAS, Hebbal, Bangalore. *Soil fertility management in organic cotton cultivation.* pp. 7-8.

ICAR, 2006. Report of the ICAR Committee on Organic Farming (*Jaivik Krishi*) Standards for *Jaivik* (organic) agriculture Products for domestic market labeling. 77p. Submitted to National Knowledge Commission. 2006.

Kairon, M.S., P.P. Tarhalkar, T.P. Rajendran, M.V. Venugopalan, and O.M. Bambawale, 1998a. Organic cotton cultivation – Technology and issues to be addressed. Journal of Indian Society for Cotton Improvement Special issue on the Natural and Colour Cotton Workshop held in May 1997.

Kairon, M.S. and P.P.Tarhalkar, T.P.Rajendran, O.M.Bambawale and M.V.Venugopalan, 1998b. *Technology generated for organic-cotton cultivation. ICAR NEWS* 4(2):1-2.

Kalisko, E. and L. Kalisko, 1939. Agriculture of Tomorrow. Kalisko Archives, London. Rudolf Hauschka. The Nature of Substances. Vincent Stuart Ltd. London.

KVM, 2004. Report on Organic Cotton Farming Workshop organised by Kheti Virasat Mission 14th & 15th May 2004, Bhai Bhagtu Jaivik Kheti Vidyalaya, Village-Bhagtuana, Near Jaitu, District Faridkot, Punjab.

Pfeiller, E. 1973. Biodynamic Farming and Gardening, Pauma Valley Company, California, Rudolf Steiner 1974. Agriculture, R.Steiner Press London.

Planning Commission, 2002. Report of the Working group on organic farming of the Planning Commission. 2002. Government of India, New Delhi.

Rajendran, T.P. and Venugopalan, M.V., 1999. Organic Cotton Farming. Chapter 24 p. p. 414-427. *In.* Handbook on Cotton in India V.Sundaram, A.K.Basu, S.S.Narayanan, K.R.Krishna Iyer and T.P.Rajendran, Eds. *Indian Society for Cotton Improvement*, Mumbai 252 pp.

Rajendran, T.P., M.V. Venugopalan, and P.P. Tarhalkar, 2000. Organic Cotton Farming in India. CICR Technical Bulletin No.1. 39 p + X. Central Institute for Cotton Research, Nagpur.

Rupela, O.P., Mehta, S. and Gowda, C.L.L. 2006. Addressing small farmers' needs through low-cost biological approaches of crop production and innovative local marketing system of farm produce - a discussion paper.

Tandon, H.L.S. (Ed.) 1994. Fertilisers, organic manures, recyclable wastes and biofertilisers – components of integrated plant nutrients. Fertiliser Development and Consultation Organisation, New Delhi pp.148+viii.

Tarhalkar, P.P., M.V.Venugopalan, T.P.Rajendran, O.M.Bambawale and M.S.Kairon 1996. Generation and evaluation of appropriate technology for organic cotton cultivation in rainfed vertisols. *Journal of Indian Society for Cotton Improvement,* 21:111-122.

Thampan, P.K.1995. Peekay Tree Crops Development Foundation, Cochin, Kerala, 354 p.

□□□

2

Scope of Organic Pest Control Options in Integrated Crop Pest Management

SHARMA[1], O.P., D.K GARG[2] & O.M. BAMBAWALE[3]
Principal Scientists Plant Pathology[1], Entomology,[2]; Director[3]
National Centre for Integrated Pest Management,
LBS Building, Pusa Campus, New Delhi-110 012

Green revolution has led the country to 'self sufficiency' in cereal production, however currently agriculture is facing second and third generation problems e.g., loss of soil fertility and texture, pollution due to agrochemicals, decreasing productivity, pest resistance and toxic residues in the food chain. During last 45 years, agro-chemicals (fertilizers and pesticides) have suddenly become an 'integral part' of Indian agriculture. This was largely a result of industrial revolution which led to large-scale manufacture of agro-chemicals, for which buyers were needed. This is where farmers came into the picture. This need was fulfilled by educating the under-developed and developing nations in that direction, influencing their policies to favour a larger consumption of agro-chemicals. Chemical pesticides while protecting crops, its injudicious use has adversely affected beneficial fauna and functioning of agro-ecosystems, leading to increased risk of secondary pests outbreak. While yields are on the decline, crop production costs has increased due to spiraling use of agro-chemicals resulting in lower farm profitability e.g., the resistance of cotton insects against the insecticides increased manifolds and compelled the use of more number of sprays, and thus created a vicious cycle, escalating cost of cultivation.

One of the options to increase food productivity lies in the improvement of traditional agriculture relying on very few external inputs. It is also called as "Jaivik "or organic farming, which is a holistic production management system. It aims to promote and enhance agro-system health, biodiversity, biological cycle and soil microbial activities. Since January 1994 "Sevagram Declaration" for promotion of organic agriculture in India, organic farming has grown manifold. A number of initiatives at Government and Non-Government level have given it a firm direction. Total area under certification process is increasing steadily. It has grown more than 10 fold from 42,000 ha during 2004-05 to more than 5.38 lakh ha in 2006-07.

Approach to organic production involves the integration of many practices (cover crops, strip cropping, grazing, crop rotation, etc.) into a larger system. Organic farming management relies on maintenance and replenishment of soil fertility and on development of biological diversity to disrupt pests and enhance natural defense mechanism to manage pests. Organic farmers do not use synthetic chemicals in any form. More than 60 % of India's arable land is under traditional agriculture, where farmers are not able to purchase costly chemical inputs and currently no synthetic inputs are being used. Likewise there are many other areas in India where crop residues, manures, legumes and neem are being used to grow healthy crops. They rely on crop rotation and intercrop, to get sustainable yields. The ecological goods and services of organic agriculture are reviewed by Shepherd *et al.* (2003). Although, the products grown under such systems may not be certified as genuine organic products, the same can be labeled as "green food" and fetch a premium price in international market under "Fair Deal". Now it is high time that attempts are made to classify these practices accordingly. If this is done, the poor farmers will get a premium price for their low yields. This will also go a long way in alleviating poverty and raising the living standards of the poor villagers. Most of the agriculture in backward and tribal areas could be safely classified as organic. We must take advantage of this opportunity by arranging a market for these products both for domestic and export market.

1. Pest Management Tactics

Inspite of the general belief that organic crops are generally "healthy", variable crop losses do occur due to insect pests and diseases. In such a situation, pests can be managed by pest management strategies which are largely preventive, rather than reactive. The non-chemical methods of pest management form the major components of organic farming. Some of the non-chemical insect pests and disease management tactics, which can be utilized in organic farming, are given below.

1.1 Cultural Control

The need for ecologically sound, effective and economical pest control methods have prompted renewed interest in cultural methods of pest control. Some of the diverse ways in which cultural methods have proved effective include sanitation (to remove sources of primary inoculum), tillage (to kill resting propagules of soil resident pathogens), application of fortified manures and soil amendments either with plant residues or bio-pesticides, habitat diversification such as crop rotation (to break pathogen life cycle), trap cropping, intercropping (to reduce inoculum), time of planting (to avoid exposure of vulnerable plant growth stage), nutrient and water management etc. The potential for carry over of pest from one season to another can be reduced by destroying stubbles, weeds and other alternate hosts of plant diseases and stem borer by ploughing the field after crop harvest and before sowing. Ploughing during summer helps in eliminating soil pests either by excessive heats or by exposing to predators. Armyworms and maize stalk borers often increase in crops grown under reduced tillage and surface accumulation of crop residues under reduced tillage conditions increases the infestation of termites, slugs, some nematodes and diseases. A diverse rotation, using legumes and other cover crops, is at the heart of good humus and biodiversity management in an organic cropping system. Crop rotation also breaks the pest cycle, reducing the incidence of insect pests and diseases. Lower incidence of insect pests was found on legumes intercropped with maize. Intercrops of spinach, beans and tomato reduced the incidence of aphids (*Brevicorynae brassicae*) and caterpillar of diamond back moth (*Plutella xylostella)* in cabbage substantially. Incidence of pod borer (*Helicoverpa armigera)* reduced in chickpea when grown in association with barley, mustard, linseed or coriander. Incidence of chickpea wilt and root rot of pea were considerably reduced when planting was delayed. Castilla (1995) has indicated that alternate wetting and drying helps in management of sheath blight disease of paddy. Maize sown late suffers little from damage by maize borer, as by then the egg parasitoid *Trichogramma* is able to keep down the population of the pest. Rice is reported to suffer less from borer attack if transplanted early (by mid June). Early maturing cotton varieties (Desi) have become popular as they escape pink bollworm attack. The scab of potato at the time of tuber formation, wet weather diseases such as anthracnose of beans; early blight and charcoal rot of potato can be checked by furrow irrigation. Diseases, which are favored by high moisture at the soil surface like damping off or collar rot can be minimized by planting crops in ridges or raised beds. Growing of pigeonpea on ridges can prevent *Phytopthora* blight, which otherwise is capable of turning as epidemic (Sharma *et al.*, 2004).

1.2 Host Plant Resistance

The use of plant varieties tolerant or resistant to pest attack is a proven, effective, economical and safe method of pest management. Plants have diverse means to avoid, tolerate or recover from attack of other organisms. At present the transgenic or genetically modified plant are not allowed in the protocols of organic cultivation, instead emphasis is on "heirloom" seeds, which over the ages through the process of selection has acquired resistance mechanism against key pests and abiotic stress. These cultivars have proven their worthiness by mean of providing sustainable yield along with high nutritional quality even during epidemics and prolonged abiotic stress.

1.3 Physical and Mechanical Control

Physical and mechanical control based on knowledge of pest behavior can help farmers to develop physical barrier or traps to keep pests away. Placing plastic- lined trenches along potato fields to trap migrating Colorado potato beetles is an example of a physical control. Physical shaking of pigeonpea plant to shake off *Helicoverpa* larvae is a common practice in parts of Andhra Pradesh and southern Maharashtra. Similarly rope running in paddy fields dislodges larvae and helps in reducing pest loads. Using the light trap in and around the fields during limited initial evening hours in soybean, pigeonpea and other crops help in capturing the adults of insect pests like Bihar hairy caterpillar, tobacco caterpillar, gram pod borer, stem fly, girdle beetle, white grub etc. Yellow sticky trap as well as yellow painted cans smeared with grease are fixed in the fields to attracts and kill whiteflies and other hopping/flying small size insects e.g., aphids & jassids. Water filled trenches in orchards traps and kills beetles and bark borers. Burning rice straw has been reported to result in reduced incidence of stem rot of rice caused by *Sclerotium oryzae*. Solarization i.e. covering soil surface with polythene sheets to increase the soil temperature has been found lethal to soil borne pathogens, hibernating insect pests, nematodes and weeds. This practice has been effective and of common practice in nurseries engaged in the business of vegetables, paddy, and plantation and flower market.

Pheromones have been proposed, evaluated and practiced as monitoring and trapping components of integrated management of insect pests for last two decades. The results have been partial in several cases whereas spectacular success has been achieved in certain cases. They are used in agricultural system, orchards, plantation or forestry for monitoring pest situation, mass trapping or mating disruption (Dubey and Sharma, 2005). Synthetic sex pheromones are commercially available and are used for surveillance, monitoring and control of many lepidopterous pests such as American bollworm, spotted bollworm, codling moth, potato tuber moth, diamond back moth, yellow stem borer etc. Mass trapping has been attempted for yellow rice borer and gram pod borer. The aggregating pheromone (Methoxy benzene) isolated from the abdominal glands of calling female beetles of *Holotrichia consanguinea* has proved very effective component of IPM in groundnut in Dausa district of Rajasthan for managing whitegrub population. Methyl eugenol dipped sponge has come as a handy pheromone trap for fruitflies in Mango orchards in northern India.

1.4 Biological Control

Biological control strategy against insect pests and diseases has been attempted through conservation, colonization and augmentation of natural enemies and introduction of exotic bioagents. The control of *Pyrilla* in sugarcane by conservation of nymphal and adult parasitoid, *Epiricania melanoleuca* is the most successful example. In UP and other adjoining states it was a common practice to spray insecticides in the month of April or May against *Pyrilla,* at the stage when the parasitoid population had just started building. After *Pyrilla* epidemic in 1985, a team constituted by the Govt. of India had recommended that if sufficient number of *E. melanoleuca* is noticed in *Pyrilla* infested area, then no chemical control is warranted. It has brought very encouraging results and since then no *Pyrilla* epidemic has been observed.

Bioagents for insect pests are available in nature abundantly (Bentur *et al*, 1994; Brar *et al*., 1994). Different state departments and private entrepreneurs have commercialized *Trichogramma*, an egg parasitoid available in the nature. The field releases of this parasitoid have been found effective against lepidopteran pests in different crops such as cotton, sugarcane, rice, vegetables etc. *Tetrastichus, Chelonus, Bracon, Goniozus, Telenomus* etc. are some of the other parasitoids attacking insect pests. Insect predators such as *Chilocorus, Cryptolaemus, Menochilus* feed on mealy bugs, coccids, scales and mites on citrus, grape vine, guava etc. The spiders and mirid predators *Cyrtorhinus* attack brown plant hopper and white back planthopper of rice. The green lacewing *Chrysopa* is another important predator of aphids, mealy bug and young caterpillars and its releases in the field have been found effective against sucking pests and bollworms in cotton in Gujarat. Some of these parasitoids/ predators can be reared in the laboratories and released. However, the period of release should synchronize with the availability of suitable stage of the pest. Several methods of conservation of natural enemies such as permitting weeds to grow on the bunds, placing of bundles of rice straw for attracting spiders, straw heaps have been found effective for maintaining predators populations.

Entomopathogens are pathogens, which are associated with insects and some other arthropods like spiders and mites in a variety of ways. They may have saprophytic, commensalistic, parasitic or pathogenic association with insect host. Several pathogens including viruses such as nuclear polyhedrosis viruses (NPV) and granulosis viruses (GV), bacteria like *Bacillus thuringiensis*, fungi like *Metarhizium anisopliae, Beauveria bassiana* and *Verticillium* sp. cause diseases in insects and destroy them. These microbial pathogens (biopesticides) are applied on the targeted pests in much the same way as chemical pesticides except that they are to be used, as preventive measures as they need time to act on. *B. thuringiensis* infecting wide range of insect pests is the most common microbial insecticide in use against the insect caterpillar pests that attack a wide variety of vegetables and other crops. Another bacterium *B. popillae* is also available commercially for the management of white grubs. The biopesticide Nuclear Polyhedrosis Virus (NPV) has a great potential for the management of *Helicoverpa* and *Spodoptera. Verticillium* as well as *Paceliomyces sp* have wide use in fields severely infested with nematode problems (Singh *et al.*, 2002).

The potentiality of suppression of collar rots (*Sclerotium. rolfsii)* on sugar beet; lentil and chickpea; damping off (*Pythium graminicolum)* on sugarcane seedlings and brinjal; *Pythium* spp. on pea, tomato and brinjal; sheath blight (*Rhizoctonia solani)* on lentil and chickpea and downy mildew on pearl millet have been demonstrated by soil and seed application of fungal antagonist, *Trichoderma* spp and their strains. Now many private entrepreneurs are carrying out the commercial production of *Trichoderma* spp., alone as well as mixture of antagonistic bacteria. Among the bacterial antagonist *Pseudomonas fluorescens* has been found effective in controlling fungal pathogens such as wilt/ root rot *Fusarium oxysporum f. sp. cubense, Pythium spp., Rhizoctonia solani, R. oryzae* and *Sclerotium rolfsii* and bacterial pathogens like *Xanthomonas citri* and *P. solanacearum* in field tests. Bacterial preparations are widely being used in organic spice cultivation in southern India.

1.5 Bio-pesticides

Biopesticides include all types of agents and products derived from biotic sources, with potential for the control of insects, plant diseases, weeds and urban insect pests. These products have been used for pest control for many centuries, but it has been only in past decade that greater attention has been directed toward the discovery and development of natural products. These products include phytochemicals, microbial control agents, antagonistic agents, genetically modified organisms, hormones, pheromones, kairomones and other agents of biological origin. Since the Rio convention and implementation of WTO rules and regulations, there has been a strong industry interest in developing "reduced risk" pesticides. Emphasis has been on registering biopesticides for use on field crops, vegetables, and fruits. The increasing interest in organically grown produce has also driven development of natural-based products registered for food crops. It is hoped that some of these new products find their way to being registered for use on ornamental plants as well. Biopesticides (both botanical and bio-organisms) are increasingly becoming popular in insect pest management programmes.

Currently, considerable amount of research is being carried out with emphasis on screening and development of plant products, phytochemicals and natural products. Fortunately our country is blessed with great diversity of plants and biodiversity of microbes and it is here that the field of homeopathic medicine relying on herbal products is relatively advanced and relied upon in contrast to other South east Asian regions. With a great wealth of knowledge on herbal medicine, it is prudent to look for bioactive materials derived from plants that may have preventive or curative properties. As many as 2,400 plant species have been reported to possess pest control properties, 1,005 species have insecticidal, 384 antifeedants, 297 repellants, 27 attractants and 31 growth inhibiting properties have been identified and their number may increase manifold (Dubey and Sharma, 2005). The most commonly exploited botanicals based pesticides are of Neem (*Azadirachta indica*), Pongamia (*Pongamia glabra*), Citrus grass (*Cymbopogon* spp) and Derris (*Madhuca indica*). Roots of the plants belonging to the genera *Derris, Lonchocarpus* and *Tephrosia* contain rotenone, which is the basis of many commercial insecticides in many countries. The seeds, leaves, pods, bark and roots of *Albizia lebbeck* (safed siris) contain caffeic acid, quercetin, which has insecticidal property. *Anacardium occidentale* (Cashewnut) has phenolic compounds in it shell oil, which is insecticidal in action. Flower extract of *Butea monosperma* (red forest) possess termiticidal property due to the chalcones present in it. *Annona squamosa* (custard apple) contains annonine in its stem, leaves, semi-ripe fruits which act as insecticide. Seed oil of *Azadirachta indica* (neem) contains azadirachtin, nimbidin, etc which shows anti-feedant, oviposition deterrent, repellant, insect growth regulator and insecticidal properties. Simple neem seed kernel extract in water (5%) has been found effective against several pests including Helicoverpa, borers, cutworm, diamond back-moth, rice brown plant hopper, rice green leaf hopper, tobacco caterpillar and several species of aphids and mites. Mahua seed kernel extract (5%) is very effective against saw-fly (*Athalia lugens*). The Root exudations or extracts of marigold (*Tagetes* sps.) work as a nematicide on plant parasitic nematodes especially root knot nematodes. A number of commercial products have seen the light of the day, and a few of them still waiting for large scale manufacturing. Majority of secondary metabolites or botanicals belong to various categories, viz. Terpenoids, Alkaloids, Glycosides, Phenols, and Tannins etc, which are capable of causing behavioural and physiological changes in pests. These active principles are being used directly against pests as crude extract or are purified and formulated as insecticides / pesticides for use on commercial basis.

Despite the recent flurry of research activity on biopesticides, there is still a great need for national, regional and international coordinated effort to develop and utilize biopesticides in broad-range of pest control systems. Research on biopesticides especially phytochemicals, antagonists and fermentation products of many bacterial and fungal agents are inadequate. In general, there is lack of communication and interaction among researchers in this field.

Among microbial biopesticides *Trichoderma* spp., is one of the fungi that are present in nearly all-agricultural soils and in decaying plant debris. In natural substrate they grow tropically toward hyphae of other fungi, coil about them in a lectin-mediated reaction, and degrade cell walls of the target fungi (Sharma, *et al.*, 2004). This process (mycoparastitism) limits growth and activity of plant pathogenic fungi. In addition to, or sometimes in conjunction with mycoparasitism, individual strains may produce antibiotics (Singh, *et al.*, 2005). A number of strains have been isolated and considerable numbers of commercial preparations are available in market for root disease management. Pathogens could be stored for three months at room temperature and six months to one year in refrigerator. Apart from these entomopathogenic fungi *Metarhizium anisopliae, Beauveria bassiana, Nemourea rileyi* etc., had been widely used against several lepidopterous pests and specific success has been achieved in case of white grub. Viral pathogens, the Gv and NPV are now commerciallly available for the management of key pests such as *Spilosoma obliqua, Amsacta albistriga Spodoptera litura, Helicoverpa armigera* (Dubey *et al.*, 2002). Commercial preparations of *Bacillus thuringiensis* have become popular for the management of *Plutella xylostella* and *H. armigera*. The bioagents and microbial pesticides (Bt., fungal, and baculoviruses) have been widely used in the management of pests and avoiding the development of resistance. Entomopathogenic fungi and bacteria paralyses or kill their hosts by adversely affecting growth and development of host insects. These pathogens are now commercially available (Dubey and Sharma, 2005).

Amongst the microbial pesticides, at least 16 microbes have been in use in the country. Out of these *Bacillus thuringiensis* kurstaki, *Trichoderma* species & NPVs are the most accepted products. 25 microbial products including *Bacillus thuringiensis, B. sphaericus, Trichoderma* species, *Beauveria bassiana* & NPVs are registered with Central Insecticides Board. The 15 parasites and predators widely used in the country do not require registration in India and only 10 are commercially available (Dubey and Sharma, 2005). The Central Insecticidal Board has accepted and included the following microbial products and chemicals in the schedule of Insecticides Act (1968).

- Entomogenous fungi e.g. *Beauveria bassiana, Verticillium lecanii, Metarrhizium anisopliae, Nomuraea rileyi* etc.,
- Antagonistic fungi and bacteria e.g. *Trichoderma* spp. , *Pseudomonas fluorescens, Bacillus subtilis, Gliocladium* etc.,
- Mycoherbicides
- Insect repellents
- NPVs and GVs
  (Source: www.cibrc.nic.in)

2. NCIPM Experience with IPM in Basmati Rice in U.P, Haryana and Uttarakhand

For the past 15 years, the government has promoted IPM as a 'golden mean' between chemical and organic farming, and has attributed declining trends of pesticide consumption to successful implementation of IPM programme. IPM is little complex than purely cultural or biological control since it requires thorough knowledge of both biological and chemical pest management, and entails combined or alternating use of chemical and biological inputs where margin for error is very small. NCIPM has been involved in field validation of IPM technology in scented rice in farmers' participatory mode in the states of UP, Haryana and Uttarakhand. The IPM technology has not only been successful in replacing chemical pesticides with microbial and bio-agents but enhanced yields with highercost benefit ratio. Studies carried out on conservation of natural enemies in IPM fields quite encouraging. Adoption studies indicate that farmers on their own are using IPM strategies (Garg, *et al.*, 2000 & 2002).

2.1 NCIPM Experience with IPM in Organically Produced Basmati Rice in Kaithal, Haryana

Field trials were carried out at Kaithal during 2005 to 2007 crop season in 50 ha farms in farmers participatory mode amidst natural infection/infestation. Trial was replicated five times for each treatment e.g., IPM, indigenous traditional knowledge (ITK) and farmers practice (FP). The working of IPM as well as effectiveness of bioagents in rice ecosystem has already been proven and same source was used in present trial. In present studies chemical pesticides have been completely replaced with commercially available alternative microbial pesticides with IMO certification as prophylactic measures. Previous studies indicate that bio-pesticides and bioagents are effective only if applied in the initial stage of pest/disease development. Fields were prepared as per standard agronomic practices and other good crop practices e.g., green manuring, Trichoderma fortified farm yard manure (FYM). Plant protection strategies adopted and compared are as follows: seed as well as seedling treatment with *Pseudomonas fluorescence*, fixing of pheromone traps for monitoring yellow stem borer (YSB), rope running at interval of 15 days, release of *Trichogramma japonicum* for yellow stem borer and leaf folder, spray of *Bacillus thuringiensis*, neem oil, spider conservation with the help of straw bundles, preventive sprays of Trichoderma, Pseudomonas, and Herbozyme was compared with the farmers using indigenous traditional knowledge (ITK).

ITK farmers practices were based on green manuring with daincha (*Sesbania rostrata*) or sunhemp (*Crotalaria juncea*), field spread with chopped greens of bio-diesel plant (*Jatropha* sp.), aak (*Calotropis* sp.), malabar (*Adathoda* sp.), FYM mixed with de-oiled mustard cake, spray of cowdung ash as desiccants, fermented crude extracts of plant (chilly (*Capsicum* spp) + garlic (*Allium sativa*); neem (*Azadirachta indica*), nirguni (*Vitex negundo*), aak, dhatura (*Datura metel*), Jatropha etc.), panchgavya formulation (ghee, buttermilk, cow urine, leaves etc.), vermiwash (10%) and homa (smoking with lantana). Conventional farmers used calendar based application of chemical pesticides e.g., cartap hydrochloride, endosulfan, monocrotophos for management of key insect pests and spray of carbendazim and hexaconazole for diseases. Comparison of pest incidence in organic fields indicates that Integrated Pest Management (IPM) strategies based on scientifically proven alternative pest management components are more effective than traditional and conventional chemical based methods. Significant difference in incidence of key insect pests and diseases indicates effectiveness of microbial and botanical pesticides against their target organisms. The yield data indicates that farmers adopting IPM practices were able to get higher yield and better cost benefit (C:B) ratio (1:5.55) in comparison to ITK (1:4.05) as well as FP (1:4.0). Quantification of soil and environmental benefits and impact of IPM practices were quite encouraging in contrast to conventionals. Results indicates that IPM resulted in better conservation of natural enemies as evident from presence of numerous predators and parasitoids like spiders, crickets (egg predators), damselflies, ants, beetles, wasps, and mermithids (Sharma *et al.*, 2006).

3. Conclusion

Several institutions and movements are making concerted efforts to promote organic agriculture in India and to bring changes in the policies favoring ecological agriculture. Through good soil and biodiversity management, farms can become increasingly self-sufficient in fertility, while pest problems are diminished. When pest incidence threatened to affect the crop substantially, biological pest and disease control measures should be resorted to. Biological methods used for pest control are plant and animal preparations e.g. neem based pesticides, pyrethrum, garlic extract, chilli extract, cow urine spray, light mineral oils e.g. kerosene, sulphur, tobacco, the viral, fungal and bacterial preparations (biopesticides) such as nuclear polyhedrosis virus (NPV), *Trichoderma, Pseudomonas* etc. These can be used when absolutely necessary and taking environmental impact into consideration. The certification agency shall be consulted before using these inputs. For disease control, naturally occurring minerals like copper sulphate and sulphur can be used in limited quantities. Use of genetically engineered organisms and products are still prohibited. Studies carried out in India by the office of evaluation of IFAD, Italy clearly indicates that "Organic agriculture" is a viable approach that can be suitable for small farm holders. Furthermore, organic agriculture efficiently reduces environmental risks by not using some potentially damaging technologies of intensive agriculture such as pesticides, herbicides, synthetic nitrogen fertilizers, GMO crops and anthelminthics. It can be particularly useful in the more difficult environment where, resources are scarce and cultivation is problematic.

References

Bentur, J.S., M.B. Kalode, B. Rajendran, and V.S Patil, 1994. Field evaluation of the egg parasitoid, *Trichogramma japonicum* Ash. (Hym: Trichogrammatidae) against the rice leaf roller, *Cnaphalocrocis medinalis* guen. (Lep: Pyralidae) in India. *Journal of Applied Entomology*, 117: 257-261.

Brar, D.S., M. Shenhmar, M.S. Mahal, Rajwant Singh and R. Singh, 1994. Egg parasitoids of yellow stem borer, *Scirpophaga incertulus* (Walker) in Punjab. *Journal of Insect Science*, 7: 61-63.

Castilla, NP, FA. S. Elazegui Savary, 1995. Inoculums efficiency in sheath blight as affected by contact frequency, leaf wetness regime and nitrogen supply. *Int. Rice Res.* Notes 20(1): 38-39.

Dubey , O.P. and O.P. Sharma. 2005. Pest management for safer environment. Pages. 65. Directorate of Information and Publication of Agriculture, ICAR, KAB-I, Pusa New Delhi-110012.

Dubey, O.P, Sharma, O. P., and K.S. Murthy, 2002. Microbial pesticides: issues and constraints . In : Biopesticide and Pest Management (eds. Opender Kaul, G.S. Dhaliwal, S.S. Marwaha and Jatinder K. Arora) pp-18-26. *Society of Biopesticides Science*. India.

Garg, D.K., M.D. Jeswani, P. Kumar, and R.N. Singh 2000. IPM approach enhances the yield of Basmati rice. *ICAR New*, 6(2) April-June: 12-13 & 15.

Garg, D.K., P. Kumar, R.N. Singh, Pathak Mahesh. 2002. Role of parasitoid Trichogramma japonicum in the management of yellow stem borer and leaf folder in Basmati Rice. *Indian journal of Entomology*, 64(2): 117-123.

Sharma, O. P., M.D. Jeswani, O.M. Bambawale, and K.S. Murthy, 2004. Trichoderma as a potential bio-agent for management of diseases: Taxonomy, biology and mode of action pp. 246-282. In: Microbial Biotechnology. Ed. Pravin Chandra Trivedi, Aavishkar Publishers, Ditributors, Jaipur.

Sharma, O.P., M. Sehgal, M.D. Jeswani, T.P. Trivedi, D.K. Garg, C.P. Srivastav, V.I. Benagi, G.L. Nigam and R.C. Lavekar, 2005. Validated IPM Technology: Pulses (B) Pigeonpea. In: Validated IPM Technologies for selected crops. (eds. Amerika Singh, H.R. Sardana and Naved Sabir). Pp. 76-90. *National Centre for Integrated Pest Management*, New Delhi.

Sharma, O.P., D.K, Garg, T.P. Trivedi, Satpal Chahar, Chouhan Promod and S.P. Singh, 2006. Effectiveness of Integrated Pest Management (IPM) strategies for organically grown export oriented scented rice. Oral Presentation (Abstract ID: 2069) Session: 2.4: Integrated Pest Management. IInd International Rice Congress 9-13 Oct., 2006 , New Delhi.

Shepherd, M.; B. Pearce, B. Cormack, L. Philipps, S. Cuttle, A. Bhogal, P. Costigan and R. Unwin, 2003. An Assessment of the Environmental Empacts of Organic agriculture. (http://www.defra.gov.uk/science/project_data/ Document Library / OF0405 / OF0405_909_TRP.doc).

Singh, Amerika, O. M. Bambawale, and O. P. Sharma. 2002. Entomopathogen bio-pesticides in Integrated Pest Management. In: Biopesticides for sustainable agriculture: Prospects and Constraints. (ed. Nutan Kaushik) pp 41-66. The Energy and Resources Institute (TERI), New Delhi.

Singh, U.S., Mishra, D.S.,Varshney, S., Zaidi, N.W., Tarana and N. Singh, 2005. Potentials and effectiveness of fungi and bacteria as biocontrol agents for plant disease management. 157-233. In: Integrated Pest Management Principles and Application. Vol-1. (Eds. Amerika Singh, O.P. Sharma and D.K. Garg) CBS Publishers. New Delhi.

❑❑❑

3

Scope of Organic Pest Control in Oilseeds Crops

VIMALA DEVI, P.S.
Principal Scientist (Agril. Entomology),
Directorate of Oilseeds Research, Rajendranagar, Hyderabad- 500 030

In the view of resurgence of interest in alternate agriculture in recent years, organic farming has been considered to be a viable option in many countries. Organic farming is a production system that avoids or largely excludes the use of synthetically produced agricultural inputs like fertilizers, pesticides, growth regulators, live stock feed additives etc. Approximately 31 million hectares (75 million acres) worldwide are now grown organically.

IFOAM (International Federation of Organic Agriculture Movements) is an international body that serves as a forum to exchange knowledge on organic agriculture. It sets and regularly revises organic standards. These standards distinguish organic from high-input farming based on their distinctively different agricultural practices. Organic systems use cultural, biological and mechanical methods instead of inputs such as chemical fertilizers or pesticides wherever possible. Any deviation from that method must be justified on the basis of need. For example, synthetic pheromone traps with substances to attract insects, have always been allowed in organic production since these traps are a sustainable tool in monitoring and controlling insect populations. According to the IFOAM principles, organic systems have demonstrated that production and processing has been possible without GMOs. Therefore, there can be no demonstrated need for GMOs in these systems. For instance, *Bacillus thuringiensis* (Bt) has for many years been used as a biopesticide in organic agriculture to control infestations by lepidopteran insects such as caterpillars. For IFOAM, organic agriculture has no need for transgenic crops, which express the same anti-lepidopteran protein.

1. Oilseed Crops

Oilseed crops constitute the second largest agricultural commodity after cereals in India occupying more than 13 % of cropped area. The major oilseed crops are groundnut, rapeseed-mustard, soybean, castor, safflower, sunflower, linseed and niger. Insect pests constitute on of the major biological constraints in oilseeds production. The economically important insect pests of oilseeds include head borer *Helicoverpa armigera*, tobacco caterpillar *Spodoptera litura*, castor semilooper *Achaea janata*, castor capsule borer *Dichocrocis punctiferalis*, red hairy caterpillar *Amsacta albistriga*, Bihar hairy caterpillar *Spilosoma obliqua*, green looper *Plusia* spp., mustard aphid *Lipaphis erysimi*, mustard sawfly *Athalia proxima*, sesame leaf webber *Antigastra catalaunalis*, sesame gall fly *Asphondylia sesame*, safflower aphid *Uroleucon carthamii* etc. There is an increasing/pressing demand for residue free food coupled with the increased awareness regarding the ill effects of indiscriminate use of chemical pesticides. The viable options include promotion of bio-intensive IPM and organic farming.

2. Pest Control in Organic Farming

Organic farming is a process which prohibits the use of synthetic fertilizers, pesticides or chemicals in growing crops. Instead, it uses beneficial naturally-occurring materials, in conjunction with proper soil and crop management practices, to develop and protect optimal soil health and crop ecology. In fact, in order to be certified organic you must be free from these materials for a minimum of three years prior to your first harvest. Organic farming results in less damage to the environment and ecosystem. From the water we drink, to the soil we farm, to the air we breathe, less contamination means a safer planet for everyone. Simply put, organic farming results in safer products for you, your family and the environment. Spraying of *Bacillus thuringiensis*, a pathogen capable of attacking several lepidopteran insects has been successfully adopted in many countries to control pests on many crops. Such bio-pesticides are developed by culturing the pathogens found in the insect species. With environmental restrictions on chemical pesticides, these alternatives are gaining popularity.

Pest control in organic farming begins by taking right decisions at right time such as growing crops that are naturally resistant to diseases and pests, or choosing sowing times that prevent pest and disease outbreaks. Careful management in both time and space of planting not only prevents pests, but also increases population of natural predators that have natural capability to control insects, diseases and weeds. Other methods generally employed for the management of pests and diseases are: clean cultivation, improving soil health to resist soil pathogens and promote plant; growth; rotating crops; encouraging natural biological agents for control of diseases, insects and weeds, using physical barriers for protection from insects, birds and animals; modifying habitat to encourage pollinators and natural enemies of pests, and using semio-chemicals such as pheromones attractants and trap pests. For reactive treatment of pest control, biological control agents are some times used in organic system. *Bacillus thuringiensis* is commonly used as a microbial insecticide against lepidopterous pests.

2.1. Practices Allowed

- Preventative management such as intercropping and crop rotations.
- Biological controls, such as release of natural predators and parasites.
- Sprays and dusts of low ecological profile, including insecticidal soaps, rock powders and diatomaceous earth.
- Microbial preparations such as *Bacillus thuringiensis*, NPV, entomopathogenic fungi etc.
- Phermones used as traps and phermones used as mating disruptive if not applied to a food crop.

2.2. Practices Prohibited

- Synthetically compounded insecticides, nematicides, acaricides, rodenticides, molluskicides, ovicides, or repellants.
- Traps containing prohibited pesticides.

- Natural poisons that are environmentally persistent and have long-term effects, such as arsenic and lead salts.

3. Microbial Pesticides

With the ever-increasing awareness of the harmful effects of the chemical on man and his environment, the immediate need for sustainable, eco-friendly pest management has been felt very strongly providing an impetus to research and development of microbial pesticides. Microbial control includes all aspects of utilization of microorganisms or their by-products for the control of pest species. Among the different microbial agents developed and tested, bacteria, viruses and fungi are considered promising for the control of insect pests (Table 1). Microbial control agents are relatively host specific and do not upset other biotic systems. They are safe to humans, vertebrates, beneficial organisms and do not cause environmental pollution and are compatible with most other control methods. They are ideal for both short and long term pest suppression. Unlike chemical pesticides, they are harmless to humans and other non-target organisms, they do not leave chemical residues on crops, are easy and safe to dispose of and do not contaminate water systems. The microbial pesticides can be locally produced and used for pest management in Organic farming. Some of the promising microbial insecticides that can be employed in organic farming are discussed below.

The microbial insecticides currently used in India in controlling economically important pests affecting agricultural and horticultural crops include *Bacillus thuringiensis* var. *kurstaki* (Btk), Nuclear Polyhedrosis Virus (NPV) of *Helicoverpa armigera* and *Spodoptera litura*, the entomopathogenic/nematicidal fungi, *Beauveria bassiana, Verticillium lecanii, Paecilomyces lilacinus* and *Metarhizium anisopliae*. By the year 2004-05, there were 13 manufacturers registered with the Central Insecticides Board, Govt. of India for manufacture of microbial insecticides (5 for NPV, 5 for Bt and 2 for *B. bassiana*) while the number has increased dramatically to 78 as on date with 65 new registrants (17 for *B. bassiana*, 20 for *V. lecanii*, 10 for *M. anisopliae* and 2 for Btk). This is a clear reflection of the increasing demand for these useful insect control agents and the immense scope for promoting organic farming in India. Organic farming employing microbials for insect pest management in India is expected to gain momentum with the assured availability of several new products of the various microbial insecticides with the increase in the number of registrants.

3.1 *Bacillus Thuringiensis*

Many different types of bacteria are known to acutely or chronically infect insects. *Bacillus* is by far the most important microbial pesticide genus. *Bacillus thuringiensis* (Bt) has been the most widely used and successful microbial pesticide ever registered. *B. thuringiensis* is an insecticidal bacterium, marketed worldwide for control of many important plant pests - mainly caterpillars of the Lepidoptera (butterflies and moths) but also for control of mosquito larvae, and simuliid blackflies that vector river blindness in Africa. Bt products represent about 1% of the total 'agrochemical' market (fungicides, herbicides and insecticides) across the world. The commercial Bt products are powders containing a mixture of dried spores and toxin crystals. They are applied to leaves or other environments where the insect larvae feed. The toxin genes have also been genetically engineered into several crop plants. Commercially available Bt preparations consist of dry bacterial spores and the crystalline toxin. Today they are used mainly in viticulture, forestry, green spaces and in potato, fruit and vegetable cultivation. They are particularly important in organic farming; Bt preparations account for around 90% of all bio-insecticides. By contrast, they comprise only one percent of agro-chemicals.

When cultured under appropriate conditions, Bt sporulates (endospore) and forms a crystalline parasporal body which contains the insecticidal protein toxin called delta endotoxin. This body is usually referred to as the crystal. Bt can be produced in a liquid medium by deep fermentation. When sporulation is complete, the bacterial cell lyses and releases the spores and crystals into the surrounding medium. The crystal is a bipyramidal aggregate of protein molecules (about 130-140 kDa) that is actually a protoxin - it must be activated before it has any effect. The crystal protein is highly insoluble in normal conditions, so it is entirely safe to humans, higher animals and most insects. However, it is solubilised in reducing conditions of high pH (above pH 9.5) - the conditions commonly found in the mid-gut of lepidopteran larvae. For this reason, Bt is a highly specific insecticidal agent. When ingested by the lepidopterous caterpillars, the crystal is activated by alkali dissolution and digestion by gut proteases, releasing the smaller, potent endotoxin of about 60kDa which binds to the midgut epithelial cells, creating pores in the cell membranes and leading to equilibration of ions. As a result, the gut is rapidly immobilised, the epithelial cells lyse, the larva stops feeding, and the gut pH is lowered by equilibration with the blood pH. The length of time until death depends on the species of insect and the dosage but feeding stops rapidly. The lowered pH in the gut enables the bacterial spores to germinate, and the bacterium can then invade the host, causing a lethal septicaemia.

B. thuringiensis and its various subspecies, strains and geographic isolates have been studied in detail for their efficacy in the field and laboratory against major lepidopterous pests such as *Helicoverpa armigera, Spodoptera litura, Leucinodes orbonalis, Plutella xylostella, Pieris brassicae, Spoladea recurvalis, Phthorimaea operculella, Achaea janata, Prays citri, Earias vittella*, etc. In India bulk of Bt is sold through Govt. subsidy programmes. The trade business is mainly on vegetables where *Plutella* on cole crops is the main target pest and *Helicoverpa* spp. on cotton, vegetables, pulses and oilseeds. Area wise segmentation of Bt products is 20% in North India, 30% in South India, 45% in Western India and 5% in Eastern India. Wockhardt has set up a hitech plant to manufacture Bt technical product and formulation based on a native strain near Aurangabad with a production capacity of 100 tonnes/ annum.

Work under the Andhra Pradesh Netherlands Biotechnology Programme at the Directorate of Oilseeds Research, Hyderabad, on Bt with specific reference to cost-effective production, field testing, transfer of technology has met with success and received the acceptance of farmers in Mahbubnagar and Nalgonda districts. A simple, cost-effective protocol based on on the principle of solid state fermentation for the multiplication of Bt has been developed for which a patent application has been filed by ICAR on 10/07/2002. One local isolate DOR Bt-1 isolated from a dead castor semilooper larva collected from farmers' fields at Kothakota has been identified for large scale field use. The formulation has been registered with the Central Insecticides Board. The technology is being offered on a non-exclusive basis for commercial exploitation. Three micro-enterprises have been established by the Directorate in A.P. in association with two NGOs.

3.2 Entomopathogenic Fungi

Concerns about negative effects of chemical insecticides have led to emphasis on alternative strategies for pest control. There is a world wide resurgence of interest in the use of entomopathogenic fungi as biological control agents, and a significant advance in development and manufacturing of these agents in the future is expected with recent biotechnological innovations.

As with the bacteria, the diversity of fungi known to infect insects is great. More than 700 species of fungi, mostly Deuteromycetes and Entomopthora from about 90 genera are pathogenic to insects. Genera that have been most intensively investigated for mycoinsecticides include *Beauveria bassiana, Metarhizium, Verticillium, Paecilomyces, Hirsutella, Erynia, Nomuraea, Aspergillus, Aschersonia, Tolypocladium* etc. The first two genera have been used on a large scale over a number of years while others have been aimed mainly at glasshouse pests or disease vectors. There are numerous biotic and abiotic constraints on the ability of fungi to infect their host insects. These include dessication, UV light, host behaviour, temperature, pathogen vigour, age etc. Many of these constraints are being addressed by advances in formulation technology.

3.2.1 Host Range

Some fungi have restricted host ranges eg. *Aschersonia aleyrodis* infects only whiteflies and soft scales while others have wide host range at the species level although individual isolates may be selective. More than 700 species of fungi, mostly Deuteromycetes and Entomophthora from about 90 genera are pathogenic to insects. Of these, Deuteromycete fungi have received most attention as they are among the easiest to produce *in vitro*. Similar long standing interest exists for *Verticillium lecanii, Aschersonia aleyrodis, Nomurea rileyi* and use of Entomophthorales. Several have a broad host range viz.,

Metarhizium anisopliae - more than 200 known hosts among Coleoptera, Lepidoptera, Orthoptera & Hemiptera.

Beauveria sp. - Identified from about 500 host species principally Lepidoptera and Coleoptera.

3.2.2 Symptoms of Fungal Infection

- Body mummified and does not disintegrate in water.
- Body covered with filamentous hyphae or mycelium.

3.2.3 Mode of Infection

Individual strains of the same fungal species often have different host ranges or pathogenicities. The infective propagule in natural Deuteromycete infections is the conidiospore which invades insects primarily thro= the cuticle. Because fungi, do not, in general have to be ingested to be infective, they have the potential (unlike Bt or viruses) for the control of sap-feeding arthropods which do not ingest pathogens on the plant surface.

The fungus invades the host cuticle through the body wall and spiracles primarily and also through the mouth parts. The conidia germinate on the insect cuticle by producing germ tubes which penetrate the body wall. The penetration is both mechanical (pressure exerted by germ tube) and enzymatic through the action of proteinase, lipase and chitinase on the cuticle. Growth of the fungus after it reaches the haemocoel is by budding which produces hyphal bodies. These are transported through out the haemocoel and give rise to localised concentration of mycelia. A heavy growth of intertwining mycelia develops in the haemocoel 1-2 days later. The larval body is completely mummified and covered by a dense white mycelial mat from which conidiophores arise close together and produce conidia 1-2 days later. Under suitable environmental conditions (essentially high humidity), death is followed by external sporulation, which helps to spread the fungus and establish an epizootic which may result in very high levels of kill. Several studies have revealed a positive correlation between RH/rainfall with rate of fungal infection thus indicating that fungal survival and spread are assured under higher rainfall and humidity conditions.

Specificity of invasion of the insect by a fungal spore may occur at several levels. The pathogen host range may first be influenced by specificity in the ability of the spores to adhere to the cuticle. Germination may be influenced not only by humidity but also by the availability of certain nutrients.

Apart from Deuteromycete fungi, the Entomophthorales (Zygomycetes) contain by far the most insect-pathogenic isolates. Optimal growth temperatures for most Deuteromycete fungi is 20-30°C making them best fit for use against tropical and sub-tropical pests. Members of the Entomophthorales appear more effective in temperate climates where they often produce extensive, if slow-acting epizootics. Most of these strains have proved difficult to culture and to store for long periods resulting in a low rate of successful field applications. Toxin production is reported for *Metarhizium anisopliae, Beauveria bassiana, Verticillium lecanii, Paecilomyces* etc. Toxin production is reported for *Metarhizium anisopliae, Beauveria bassiana, Verticillium lecanii, Paecilomyces* etc. Several studies have revealed positive correlations between RH/rainfall with rate of fungal infections thus indicating that fungal survival and spread are assured under higher rainfall and humidity conditions.

Entomofungal pathogens recorded from India include *Beauveria bassiana, Metarhizium anisopliae, Verticillium lecanii, V. aphidicola, Nomuraea rileyi* from *H. armigera, S. litura* and *Trichoplusia* spp., *Hirsutella* spp., *Erynia neoaphidis, Paecilomyces farinosus, Pandora delphacis, Entomophthora aulicae.* Several species of entomopathogenic fungi such as *Beauveria bassiana, Metarhizium anisopliae, Nomuraea rileyi, Hirsutella thompsonii* and *Verticillium lecanii* have been studied in detail for their efficacy in the laboratory and the field. Mass production techniques have been standardized. The entomopathogenic fungi have given promising results in humid and mild climate areas and seasons. In India, work with white halo fungus *V. lecanii* has brought down the coffee green bug Coccus viridis to almost negligible proportions in the Pulney hills. The green muscardine fungus *Metarhizium anisopliae* has also been useful for the control of rhinoceros beetle of coconut *Oryctes rhinoceros.* At the Directorate of Oilseeds Research (DOR), studies with the fungal pathogen *Nomuraea rileyi* have generated the basic data pertaining to effective field dose, field peristence, instar susceptibility, identification of virulent isolates, cost-effective mass multiplication and formulation. This fungal pathogen infects a majority of noctuids viz., *Helicoverpa armigera, Spodoptera litura, Plusia* spp. etc. UAS Dharwad has undertaken mass production of the fungus and is making it available to farmers. Commercial formulations of *B. bassiana* are currently available in India. Protocol for low cost mass production of *B. bassiana* as well as its formulation in oil has been developed at DOR.

3.3 Baculoviruses

Viruses of several types are known to infect insects but baculoviruses are found only in arthropods and have not been recorded from vertebrates/plants, safe to man and non-target organisms and hence have received priority consideration for development as insecticides. Insect baculoviruses are important natural regulating factors for insect populations particularly Lepidoptera. Their commercial development has been limited by the need to produce them *in vivo* and by their generally high specificity, which while desirable from an environmental point of view limits markets. Although baculoviruses have not had the commercial successes of *Bt,* they have significant potential for use in organic farming because they do not affect non-target insects (parasites and predators), safe to humans and environment. In pest management, it is important to have a selection of control agents when designing strategies. Viruses are not likely to evoke cross resistance to chemicals or to each other, so more attention must be given to viral pesticides.

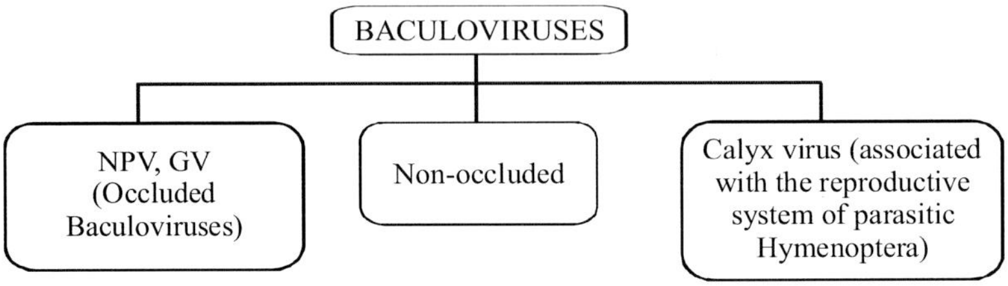

Epizootics of baculovirus diseases are frequent in lepidoptera and sawflies with very high larval mortalities resulting in strong population depression.

Epizootic Development Requires

a) Several host generations
b) High virus infecting capacity
c) Persistence of the virus inside and outside the host.

Baculoviruses survive long periods in soil. Reservoirs of baculovirus in soil have long term importance and initiate epizootics when insect populations resurge following a phase of low density.

3.3.1 Symptoms of Viral Infection

- Dead larvae may be found hanging from or lying on leaf or plant surfaces.
- The cuticle may be very fragile, rupturing easily when touched, releasing the body contents which have become liquefied.
- Porcelain white masses of fat body may be visible through the cuticle.

Baculoviridae are characterised by the presence of proteinaceous inclusion bodies within which either a single virus particle (GV) or many virus particles (NPV) are embedded. The matrix protein within which the virus particles (Virions) are embedded are called polyhedrin for NPV and granulin for GV. The polyhedral inclusion bodies (PIB) of NPV are upto 5Fm in diameter and may contain several hundred virus particles. Each virus particle may contain one or more than one nucleocapsid. The inclusion bodies of GV are called capsules or granules and measure 400-500 nm x 200-300nm. Each capsule contains a single virus particle.

3.3.2 Mode of Infection

Infection occurs most commonly via the midgut following ingestion of inclusion bodies. When infected, and under the alkaline conditions of the lepidopterous and sawfly midgut, the inclusions breakdown releasing virus particles. In lepidoptera, the virus particles pass through the midgut cells, sometimes with a replicative phase and enter the haemocoel. Major sites of replication are the nuclei of fat body, haemocytes and hypodermis.

Table 1: Candidate microbial pesticides available for pest control

Pathogens	Major Target Groups of Insects
Bacillus thuringiensis (Cry I A-G, Cry II A-C, Cry III A-D, Cry IV A-D, Cry II A)	Lepidoptera, Coleoptera, Diptera
Bacillus sphaericus	Diptera
Nuclear Polyhedrosis Viruses	Lepidoptera
Granulosis Viruses	Hymenoptera
Beauveria bassiana and *Metarhizium anisopliae*	Coleoptera, Lepidoptera, Hymenoptera, Orthoptera
Beauveria brongniartii	Coleoptera, Homoptera, Lepidoptera, Diptera
Nomuraea rileyi	Lepidoptera
Verticillium lecanii	Homoptera, Thysanoptera, Diptera
Lagenidium giganteum	Mosquito larvae
Paecilomyces fumosoroseus	Coleoptera, Homoptera, nematodes

All viral insecticides are currently produced in susceptible host insect larvae, most of which are reared on artificial diet. This is usually labor intensive and expensive but does not require much capital investment. Therefore, cost of crop protection basically depends on the number of larvae which have to be reared, infected and harvested to treat one hectare typically 100 - 1000. Labor is usually cheap and plentiful in developing countries and hence virus can be easily produced at a reasonable cost. Long term benefits can be achieved through the use of NPV since most of the dead larvae remain on the plant with their integument ruptured resulting in release of NPV laden haemolymph to persist in the soil and result in epizootic spread of the disease in the next crop.

3.3.3 Mass Multiplication of Baculovirus

Infected larvae
↓
Suspend in sterile distilled water (7 days)
↓
Decant supernatant
↓
Filter through double layered muslin cloth
↓
Suspend in distilled water
↓
Purify by centrifugation

Centrifugation at 2000 RPM for 5 min to sediment impurities
↓
Centrifuge supernatant at 10,000 rpm
↓
GV

Low speed centrifugation at at 250-500 RPM for 5-10 min.
↓
High speed centrifugation of supernatant at 3000 rpm for 30-60 min.
↓
Resuspend pellet in distilled water and Repeat alternate cycles of centrifugation
↓
NPV

❑❑❑

4

Eco-Friendly Management of Soil - Borne Pests

RAJAGOPAL, D.
Retd.Director of Instruction (PGS)
University of Agricultural Sciences, G.V.K.Campus, Bangalore-560 065.

A variety of invertebrates that inhabit the soil and participate in different activities which will result in several changes in the ecosystem. Thus the soil is a permanent habitat for some insects and other invertebrates to spend either complete or one or two stages of their life cycle. Soil offers protection from predators and parasites for many pests either in egg or larval or pupal stages of many insect pests and also spend their diapause in the soil to overcome the adverse climatic conditions.

Therefore the life cycle of the pests has to be understood before applying any management practices including organically used products. Plants are the richest source of bioactive organic chemicals on earth. They are store houses of bioactive phytochemicals such as alkalloids,s terpenoids, polyacetylenes, unsaturated isobutylamide, rotenoids etc, which may act as toxicants, repellents and modifying the behavior of insects. The usage of plant products is being encouraged all over the world to manage the crop pests including soil borne pests also.

1. Nematodes

The soil inhabiting nematodes are free living plant parasitic forms have variety of food habits such as phytophagous, predaceous and also feeding on algae, fungi, bacteria and organic matter. In India the annual loss due to the damage of nematodes is estimated to several crores of rupees in cultivated field crops. The nematode infested plant showing the symptoms are reduced growth, discolouration of foliage and stunted growth of plants. In addition to these symptoms, they exhibit reduced root growth, gall formation and subsequently affect crop yields. Majority of the plant parasitising nematodes are soil borne pests of various crops. These nematodes that live completely inside the roots and spend most of their life in soil include, the root lesion nematodes *Pratylenchus* spp, the root knot nematode, *Meloidogyne* spp, the cyst nematode, *Heteroderus* spp, the rice root nematode, *Hirshmanniella oryzae*, the reniform nematode, *Rotylenchus reniformis*, the citrus nematode, *Tylenchulus semipenetrans*. These species are known to damage the roots, bulbs, rhizomes and gall formation on the roots of vegetables.

1.1 Management

Since the nematodes are soil borne, several cultivation practices have to be modified such as crop rotations include growing non host crops to reduce the nematode population. Avoid growing solanaceous crops continuously one after another.

The most economical and effective management of nematodes is growing resistant varieties of crops to the nematodes attack. Varieties like NTDR-1, Arkavaradan in tomato have been recommended by the research workers. The cultural practices like trash burning, field sanitation, use of nematode free planting material growing cover crops or trap crops or inter crops to reduce the nematode population. Certain plants produce exudates from their roots and these are antagonistic or toxic to nematodes. The root exudates of *Tagetes spp., Asparagus sp.* and *Eragrostis sp.* will be toxic to the root damaging nematodes.

Hot water treatment of seeds and other planting material infested with nematodes. Quarantine regulations have been existing in nematode endemic areas like golden nematode infestation of potato in Nilgiris of Tamil Nadu where the infested tubers should not be transported to other states. Organic manuring in the form of oil cakes such as neem, pongamia, castor cake etc. @ 250g/sqm areas in heavily infested vegetable crops, 100g/plant in case of Banana, pomegranate, grapes, citrus and papaya. Incorporating these cakes @ 10kgs/acre for most of the field crops including vegetables and fruit crops.

2. Snails and Slugs

The phylum Mollusca are distributed throughout the world inhabiting aquatic and terrestrial ecosystems. They are heribivorous, carnivorous and fungivorous. The snails and slugs are more or less similar in body structure except that the slugs lack in the external hard spiral protective shell and their mantle is smoother. Both of them withdraw their body when disturbed. On many occasions, snails and slugs become serious pests of many cultivated crops. They gnaw large irregular holes on foliage, tender stems of seedlings, tubers and roots. Majority of them are nocturnal in behavior (Rajagopal, 1988a).

In India, the following species are found damaging the different ornamentals vegetable crops, mulberry, plantation crops like Banana, Arecanut, coffee, etc.

- Giant African Snail, *Achatina fulica*
- Field snail, *Ariophanta solata*
- Garden Snail, *Opeas gracile*
- Black slug, *Laevicaulis alte*
- Brown slug, *Mariaella dussumiere*

The Giant African Snail, *Achatina fulica* is an unusually big sized snail measuring 15-20cm in length. Each snail lays about 50-100 eggs once a year in groups either on the soil surface or just below the soil surface. The eggs are quite conspicuous yellowish white, spherical in shape 4 -5mm in diameter. Incubation period varied from 12-15 days under Karnataka conditions. The life span is quite long varied from 3-5 years. These are very active during rainy season. During unfavorable weather conditions of winter or summer months they remain inactive.

The black slug *Laevicaulis alte* is widely distributed tropical species. These slugs are black in colour, soft bodied about 8-10 cm in length when expanded. There is a distinct head with a pair of antennae and a pair of stalked eyes. It leaves of silvery shining slimy mucous trail on the surface along the course of its movement. Eggs are deposited in mass during rainy season, 4-5cm below soil or litter surface in shady places, where moist condition in the habitat prevails. Each mass of eggs resembles a beaded chain and contain 38-80 eggs are oval translucent and dark grey with an elastic outer membrane. It hatches in about 8-10 days. Young ones feed on the leaves by making small punctures and adults feed on the leaves voraciously and caused defoliation. They were also found feeding on the fruits of vegetables.

2.1 Management

- Cultural practices like removal of weeds and keeping the garden and surrounding area clean and removal of stubbles or bushy vegetation in the fields are the important aspects of reducing the pest population.
- Mechanical collection adult slugs and snails in the crop fields and also when they congregated during their hibernation or aestivation. During off season, the hiding places can be located and destroy them.
- Many biocontrol agents such as predatory snails, toads, crabs, carabid beetles, ants, domestic ducks, have been recorded as predators of eggs, young ones and adult snails and slugs.
- Dusting of slaked lime, powdered tobacco, common salt to the breeding places and foraging tracks in the field will be effective in managing the snails and slugs.

3. Termites

Termites are highly organized social insects living in small to large colonies. They are polymorphic, exhibit division of labour, morphologically different as male (King) and female (Queen), workers and soldiers and these are living in the colony as reproductive castes and sterile castes respectively. The close association of termites especially with field crops, fruit crops forest plantations and woodwork of the buildings has attracted considerable attention primarily, because of the heavy damage they cause. In addition, they are also known to feed on the organic matter like leaf litter, root stubble and dung in the ecosystem. Reports have been indicated from different parts of the world, they are eating the valuable documents and damaging monuments preserved in the historical museums and also damaging ancient buildings (Rajagopal, 1988b).

There are two types of termite habitats: 1. Soil inhabiting 2. Wood inhabiting. The soil inhabiting termites are quite common and dominant as soil fauna either as mound or nest builders in the ground. Wood inhabiting termites are arboreal types which construct their nests either inside or outside the trees. Among the different castes of termite colony only the worker population that does the damage to crops but queen is the center of activity without killing her, there is no real control of termites. Termites constitute important group of soil fauna and play a major role in the arid and semiarid tropics. Worker castes are dominant and are responsible for major functions in the colony during the process of food collection, the worker castes generally attack the plants that are weakened due to some internal growth factors or environmental causes such as drought or poor soil conditions.

3.1 Management

Destruction of mounds and killing the queen termite is the real control of termite population. The natural control using predators, pathogens and plant products such as neem and caster cakes apart from these several cultivation practices which can be used effectively in different cropping systems for managing termites. Frequent irrigation helps in reducing the attack by *Microtermes obesi* in crops like maize, wheat, sugarcane, groundnut (Verma *et al.*, 1981). Groundnut pods dried by placing on the neem cake or *Ipomoea fistulosa* mulches in groundnut cropping system reduce 80-90% termite damage to pods. Application of well decomposed farm yard manure to maize crop reduced termite damage compared to partially decomposed farm yard manure. Continuous ploughing with frequent weeding and inter cultivation disturb the termite activity and reduces the damage to the crops.

Termites are more vulnerable to natural enemies. Ants like *Leptogenys processionalis, Oecophylla smaragdina* have been observed to predate the termites. Other predators such as lizards, frogs, birds, carabid beetles etc. are found feeding on termites in the nature.

Verma *et al.* (2004) investigated on the efficacy of *Ageratum conyzoides* and *Lantana camera* of 5% leaf extracts have been exhibiting more promising insecticidal activities against worker termites of *Microcerotermes beesoni* which exhibited 65-70 and 63 to 68 percent mortality of workers within 48 hours respectively under laboratory conditions.

The clearing and burning of crop residues and other vegetable debris deprives the food for subterranean termites and helps to keep their population at a low level.

4. Ants

Ants have been nuisance in the dwelling places and also in many agricultural fields. They are omnivorous, grainivorous, phytophagous, fungivorous and carnivorous in nature. Majority of the ant species have close association with aphids, scale insects, mealybugs, hoppers etc.for their honeydew excretions they are sporadic pests of many agricultural and horticultural crops. They are often trouble some in the agricultural fields and plantations, where they interfere with working people by their sharp bites.

The following species are commonly seen as pests of different crops.

Sl. No.	Species	Hosts
1.	*Solenopsis geminata*	Germinating seeds of cereals vegetables like brinjal, okra, cruciferous etc.
2.	*Dorylus labiatus*	Potato, groundnut mustard, mango, tuber crops.
3.	*Camponotus compressus*	Cereal and millet crops
4.	*Anoplolepis longipes*	Stored products
5.	*Oecophylla smaragdina*	Orchard crops like mango, sapota, guava etc.
6.	*Pheidole sp.*	Wheat, Rice and maize
7.	*Atta cephalotes*	Coffee, citrus, cocoa etc.

4.1 Management

Ants can be managed by using organically based attractants and repellents such as methoprene and turmeric based compounds.

Freshly ground turmeric is reported to act as a repellent to most of the ant species such as *Monomorium, Solenopsis* and *Tapinoma* (Viswanath and Veeresh, 1988).

5. White Grubs

Among soil borne insect pests, white grubs form a major component both in number of species and diversity of damaging the different crops in the Indian subcontinent. With the advent of high yielding verieties and changed pattern of cultivation, the white grubs have assumed a serious threat to cultivated crops in India and caused heavy losses and hence they are considered as one of the major national pests in India. Adults of the white grubs are commonly known as "chafers or May-June beetles" and their larvae are root feeders hence they are known as rootgrubs of polyphogous nature as they are attacking all types of crops except the crops grown in standing water.

The beetles emerge after the first rains in summer during April-May at dusk between 19.00 to 20.30 hrs. They are attracted to certain plants like neem, guava, grape, rose, moringa, pithacalobium, acacia etc on which they settle and mate immediately after emergence. They return back to the soil before dawn and lay eggs in the soil for 2-3 weeks. Eggs hatch in 10-15 days. The first instar larva feed on organic matter for a month. The second and third instar larva feed on roots for 3-4 months from June to October and some times may extend till January. They pupate in September to January depending on the development of larva. The pupal period lasts for about 10-12 days but the adults after formation remain in the soil till the next rain (Veeresh, 1988).

5.1 Management

- Many biotic factors like parasites, pathogens and predators will reduce the population in the nature. Continuous flooding for more than a week and prolonged drought at the time of adult emergence are the important abiotic factors causing natural decline of root-grub population.
- Organise adults collection at the time of emergence during the dusk and killing them after the first rains in summer for a week in the evening hours between 19.00hrs to 21.00hrs using petromax lights to locate and collect the beetles.
- Continuous hoeing and repeated ploughings 4-6 times after harvest of the crops exposing the grubs and pupae instantaneously predating by wide variety in the nature.
- Creating the puddling condition of the field in standing water for 8-10 days will kill all the stages of the white grubs.
- Fungi such as *Metarrhizium anisopliae, Beauveria tenella, B. bassiana* and *B. brongniartii* have been reported to occur as pathogenic to the rootgrubs from different parts of India.

6. Weevils and Flea Beetles

The genus *Myllocerus* includes ash grey weevils and they are ployphagous in nature. More than 78 species are reported from India and these are widely distributed in the world. The common species of *Myllocerus* distributed in South India, are *M. discolor M. viridanus, M. dentifer, M. maculosus* and *M. subfasciatus*. Both grubs and adults are phytophagous feeding on variety of crop plants such as cereals, millets, pulses, oilseeds, jute crops, vegetables forest plants etc. The grubs are feeding on the rootlets and scrape taproots up to collar region cause wilting and ultimate death of plants. Adults are defoliating the crop plants and all the stages of the Juveniles are found in the soil. All the species of *Myllocerus* are confined mostly in and around cultivated plants. Adults lay single or groups of eggs in the soil. The grubs are feeding on the root system and also collar region of the plants. Pupation occurs in soil near the root system. The adults are invariably found to feed on the foliage after emergence from the soil and continue to reproduce throughout the year. Apart from *Myllocerus* spp. *Cosmopolitus sordidus* on banana rhizome, *Cylas formicarius* on sweet potato are found to be common as soil inhabiting pests of the crops in India. The pumpkin beetles (*Raphidopalpa* spp *and Aulacophora* spp.) are commonly called as flea beetles inflict serious damage to cotyledonary leaves of cucurbitaceous crops such as pumpkin, gourds, melons, cucumber etc. The adult beetle makes punctures on the growing shoots and leaves. The grubs are known to damage the roots and other plant parts touching the ground.

A number of bio control agents such as carabid beetles, spiders, lizards, birds have been reported to feed on grubs and adults of weevils.

7. Mealy Bugs

Soil inhabiting mealybugs have been associated with roots of field crops and also fruit crops, orchards etc. The pineapple mealybug, *Dysmicoccus brevipes* suck the plant sap by congregating around the roots and the basal portion of the leaves and fruits in pineapple.

The mealybug population under the subterranean habitat is more injurious mainly because of their injury as well as acting as vector of the pineapple wilt disease. The symptoms of damage caused by mealybugs are discoloration of leaves which turn reddish yellow with stunted growth, ultimately reduced size of fruits and death of plants. Similarly the nymphs and adults of root mealybug, *Planococcoids robustus* are congregating on the roots of mango crop especially in the early stage of its growth. As a result, the plants turned to yellow and started wilting. The roots are found to be coated with black sooty structure due to honeydew excretion which results in rotting of underground roots and drying of plants (Rajagopal and Govindan, 1988).

8. Cutworms

They are considered as important soil dwelling Lepidoptera, majority of the cutworms are taking shelter in the soil during day time and attack the crops during night time. Pupation takes place inside the soil. More than 10 species of cutworms including *Spodoptera litura, Agrotis ipsilon* and *Laphygma exigua* are known to have close association with soil with polyphagous in nature on important cultivated crops all over India. They are commonly seen on potato, tobacco, cotton, pulses, groundnut etc (Rajagopal and Govindan, 1988).

8.1 Management

Few species of cutworm moths are known to attract the light. The light traps and sex pheromone traps are useful in monitoring the pest population.

Egg mass and early stage of the congregating caterpillars on the hosts can be collected and destroy them.

Apanetles ruficus a braconid, has been recorded as an important larval parasite on cutworms. Many plant products such as neem oil, pongamia oil have been advocated to manage the cutworm population.

References

Rajagopal, D., 1988a. Snails and Slugs pp. 110-129. In: *Applied Soil Biology and Ecology,* II edition, Oxford and IBH Publishers, New Delhi.

Rajagopal, D., 1988b. Termites. pp. 180-218. In: *Applied Soil Biology and Ecology,* II edition, Oxford and IBH publishers New Delhi.

Rajagopal, D. and R. Govindan, 1988. Other Soil Insects. pp.283-316. In: *Applied Soil Biology and Ecology.* II edition, Oxford and IBH publishers, New Delhi.

Veeresh, G..K., 1988. White grubs. pp. 248-292 In: *Applied Soil Biology and Ecology,* II edition, Oxford and IBH publishers, New Delhi.

Verma A.N., A.D. Khurana, and B.S. Malil, 1981. Effect of different levels of irrigations and fertilizers on termite damage and yield of wheat crop. In: *Progress in Soil Biology and Ecology in India* Ed. G.K. Veeresh, *UAS, Tech. Ser.* 37. 285-291.

Verma, S.K., R.K. Verma, and K.D. Saxena, 2004. Bio efficacy of herbal extracts against building termite, *Microcerotermes beesoni*. *Pestology*. 28(6)19-22.

Viswanath, B.N. and G.K. Veeresh, 1988. Ants. pp. 219-242 In: *Applied Soil Biology and Ecology* II edition, Oxford and IBH publications, New Delhi.

❑❑❑

5

Organic Pest Management in Agricultural Ecosystems

NADARAJAN, L.
Principal / Co-ordinator, Krishi Vigyana Kendra, Karaikal,
Union territory of Puducherry 609608

1. Introduction

Modern developments have made the scientists, administrators and planners to think seriously to make a somersault backing to the old day practice i.e., organic farming. The use of 'Panchakavya' and organic pest managements are showing reentry into the farming system. However, the increasing needs of the people will have to be met by boosting productivity and this in turn means that quality crop production and protection inputs will have to be used. In India, it has been estimated that, the total crop losses due to pests accounts 90,000 crore rupees. The use of chemical pesticides remains one the best known and most widely used pest control tactics. It was World War II that opened the chemical Era with the introduction of a totally new concept of insect control using synthetic organic insecticides, the first of which was DDT.

2. Pesticide Scenario in India

About 144 Pesticide molecules are registered in India. A total of 65 technical grade pesticides are manufactured indigenously. There are around 400 manufacturing units involved in the production of technical pesticides and their formulations. Almost every MNC have invested in the production of pesticides in India. Certain Indian Companies have made substantial investments and have made a name and reputation for themselves in the world markets. India is now recognized as an important source for supply of generic products. A very large number of units in the small scale sector are involved in formulations and sell their products essentially on regional basis. It would seem that India should be in a very happy situation in protecting its agricultural wealth from the ravages of pests and diseases. Certain environmentalists have even opined that large scale use of pesticides is seriously affecting our environment. Yet we lose more than Rs.20,000 crore rupees of crops due to pests. The per hectare consumption in India is a paltry 570 g/ha against 2500 g/ha in USA., 3000 gr./ha in Europe and 12000 g/ha in Japan. Our consumption is even lower than the consumption in Mexico which is 570 gr./ha.

The first report of poisoning due to pesticides in India came from Kerala in 1958 where, over 100 people died after consuming wheat flour contaminated with parathion. Subsequently several cases of pesticide-poisoning including the Bhopal disaster have been reported. Despite the fact that the consumption of pesticides in India is still very low, about 0.5 kg/ha of pesticides against 6.60 and 12.0 kg/ha in Korea and Japan, respectively, there has been a widespread contamination of food commodities with pesticide residues, basically due to *non-judicious use of pesticides.* In India, 51% of food commodities are contaminated with pesticide residues and out of these, 20% have pesticides residues above the maximum residue level values on a worldwide basis. It has been observed that their long-term, low-dose exposure are increasingly linked to human health effects such as immune-suppression, hormone disruption, diminished intelligence, reproductive abnormalities, and cancer. The pesticide consumption in India over the past ten years is given below.

Pesticide usage for the cultivation of food crops among the different states of India showed a mixed pattern. Tamil Nadu consumes 1.2 to 2 kg/ha of pesticide which is followed by Andhra Pradesh and Punjab where 0.8 to 1.2 kg/ha is the rate of consumption, Gujarat, Jammu & Kashmir show pesticide consumption at the rate of 0.6 to 0.8 kg/ha, whereas states like Kerala, West Bengal and Assam use 0.4 to 0.6 kg/ha for plant protection. Interestingly, the rate of pesticide consumption is almost the same (0.3 to 0.4 kg/ha) in the northern states such as

Rajasthan, Himachal Pradesh, Haryana, Uttar Pradesh and Bihar, Meghalaya in north east and Karnataka in the south. Similarly, Maharashtra, Madhya Pradesh and Orissa consume 0.2 to 0.3 kg/ha of pesticide for crop protection. Information is not available regarding the status of pesticide consumption for Sikkim, Nagaland, Manipur and Tripura. Pesticide consumption in Arunachal Pradesh and Mizoram is 32 and 21.5 million tons, respectively in the entire states, as per figures available from the Toxicological Atlas of India prepared by ITRC/CSIR (121).

The figure reveals wide variations in pesticide consumption in various states of India as per the Toxicological Atlas of India for Pesticides (121). The environmental load of pesticides in India in term of kg/Sq.km. land area presents an interesting picture. Maximum pesticide load is found in the environment of Punjab, Haryana, Delhi and Tamil Nadu.

In this regard, emphasis in crop protection in the last decade has been on the discovery and development of potent plant protection products that are least disruptive to the environment. The focus has been made on the development of botanical insecticides with potent against phytophagous insects. Also organic pest management is emphasized to produce crop without pesticide residues to save people from the ill effects of insecticides.

3. Organic Agriculture

Organic agriculture is a viable alternative because it enlivens the soil, strengthens the natural resources ad helps in biological activity of the soil. Vedic Agriculture are the time tested and time honored agricultural practices mentioned in Rig-Veda, Krishiprasara, Manusmriti, Agni Purana and Vriksha Ayurveda. The concept of Vedic Agriculture is based on the unique philosophy of "harmony" with nature - to have respect and gratitude for animate and inanimate objects, harmony with natural forces (Pancha mahabuthas or five elements viz., earth, water, air, fire and sky) and recycling of matter. The emphasis is on saha-astitva (co-existence and co-relation with living and non-living entities) and never on dominance or exploitation. In plants and animals, a symbiotic relationship exists between matter and energy. Matter nourishes and energy protects, thus the equilibrium is maintained. Detailed information is provided in the Vedic texts. The symbiosis of energy and matter in based on this concept only organic agriculture developed from time immemorial in our great country and due to the initial interest in alleviating the poverty and increase in the production of the food grains the usage of insecticides were increasing. This can be understood by referring to the figure below

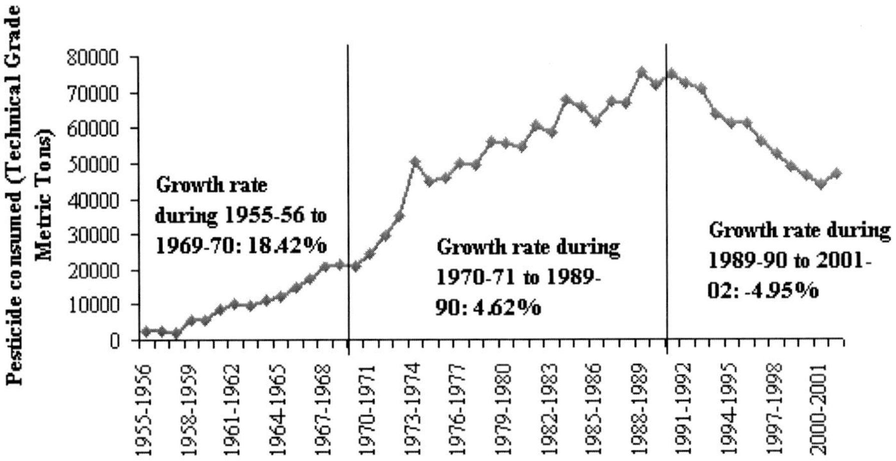

Fig. 1 : Consumption of Pesticides in India

Five elements or Panchamahabuthas	**Functions**	**Plant parts involved**
Earth or Prithvi	Controls Root functions Extraction of nutrients Development of different plant parts	Roots Stems
Water or Jala	Transports nutrients	Leaves (Veins and Petioles)
Fire or Teja	Photosynthesis Metabolism	Seeds Fruits
Air or Vayu	Respiration	Leaves
Space or Akash	Makes space available for performing the above functions	Vacoules

Now the reverse centre is noted and the farmers are shifting to IPM and also there are increase in the adoption of IPM practices and increased usage of organics/botanical insecticides and purchasing preference of organic products among the consumers and importers.

4. Alternative Approach

Such an alternative non-pesticide approach recognizes the importance of the following:

- That natural insect balances in a farm are important to control what we consider as "pests". For this to happen, the fields cannot be in a toxic-contaminated state
- That pest life cycles have to be understood and pest management has to begin right from the beginning before the eggs are laid. Several steps along the way at each stage are needed. This understanding includes close pest surveillance and decisions based on the incidence · that crop diversity plays an important role in pest management; in that sense, seeds play an important role and therefore, control over seeds by the farm communities. Trap crops and repellent crops have a role to play too
- That local and naturally-occurring materials can be used for pest control; this will also have its own political-economy dimensions which are of benefit to the farmers that since many of the pests are polyphagus, these pest management principles have to be applied across different crops and at a particular scale, for maximum benefits
- That soil nutrient management in organic ways plays a crucial role in the plant[1]s ability to withstand pest and disease incidence
- That a new paradigm of pest management can not only benefit the farmers economically and ecologically but can also address certain developmental and social issues including gender
- That such pest management need not result in decreased yields, as it is usually made out to be
- That such pest management principles and practices are pretty often drawn from farmers' experiential knowledge.

4.1 Physical Methods of Control

4.1.1 Net Barrier for the Avoidance of the Pest

The per cent damaged fruits were lowest in treatments with net barrier + weekly removal of damaged fruits and shoots (NB+REM) followed by net barrier alone as reported.

4.1.2 Sticky Traps

Yellow sticky-traps can be used to reduce the adult populations of cucumber beetle in the early stages of infestations. The addition of kairamones to yellow sticky-traps increases their effectiveness. For best results, place the traps along borders or place yellow sticky-tape above a row of affected crops. The maximum trapping of adult whiteflies were observed in the early morning because adult whitefly generally emerged in the early morning and required a minimal light intensity to become active.

4.2. Mechanical Methods of Control

The reduction or suppression of insect pest population by means of manual devices referred as mechanical control is considered to be one of the approaches for pest management.

4.2.1 Collection and Destruction of Affected Plant Parts

Collection and destruction of fallen infested fruits was effective against fruit flies and fruit borers. Treatment by mechanical destruction of infested shoots and fruits with larvae, neem oil (0.2%) and release of the egg parasitoid, *Trichogramma japonicum* Ashmead resulted in very good control of shoot and fruit borer of brinjal as compared to control.

4.2.2 Erection of Bird Perches

Preetha and Nadarajan (2002) reported that erection of bird perches as one of the component in IPM modules was found to reduce the population of semiloopers (*Xanthodes graelsii* Feisth, *Tarache nitidula* Fab., *Anomis flava), Spodoptera litura* and other lepidopteran pests in bhendi ecosystem.

4.3 Cultural Methods of Control

Cultural method of control is the deliberate manipulation of the environment to make it less favorable for pests by disrupting their reproductive cycles, eliminating their food or making it more favorable for natural enemies.

4.3.1 Effect of Sprinkler Irrigation on the Pest Incidence

The effect of sprinkler irrigation in reducing the brinjal pests was reported by few authors. The falling water drops in sprinkler irrigation was found to disturb the adult moths and drive them away, or seriously disturb their overall flying and mating activity, since it was done close to dusk when the moths were active for mating. This had an adverse effect on the larval feeding and also probably affected the aphids and thrips to a lesser extent. In addition to the reduced insect infestation, this had increased the efficacy of water utilization (AVRDC, 1984).

The controlling effect of sprinkler irrigation on diamond back moth infestations of *watercress* was due to the reduction in the egg deposition. Infestation of diamondback moth was reduced by rain compared the disruptive effect when the sprinkler irrigation coincided with physiological or behavioral periodicity with that of the demonstration for adult emerged from the pupa stage. The number of diamond back moth larvae and pupae on cabbage under overhead sprinkler were significantly less than in drip irrigated plots. The sprinkler irrigation also reduced the number of eggs laid by this pest. Significant reduction in whitefly eggs and nymphs in sprinkler irrigated crops compared with furrow irrigated cantaloupe (*Cucumis melo* L.).

4.3.2 Effect of Intercropping on the Suppression of the Pest

Intercropping was an important component of small farm agriculture in the tropics and one of the reasons for the evolution of these cropping patterns was to reduce the incidence of insect pests.

Crop rotation, an important management tool for disease control, is ineffective in controlling cucumber beetles, since the beetles migrate from areas surrounding the fields. Also, since these insects can survive on a number of wild hosts, the removal of alternative hosts from the farm would be difficult and again ineffective because of in-migration.

Cultivation and residue removal can help reduce over wintering populations of cucumber beetles. To remove infested root residues, plough the soil deeply to bring these damaged residues to the surface, then remove and destroy them.

Heavy mulching can deter cucumber beetles from laying eggs in the ground near plant stems and may hinder feeding by larvae migrating to fruits. This cultural control method, however, does not protect the leaves against attack from adult insects. Injury to fruit by tunneling of larvae is dependent on very moist soil as fruits ripen. Limiting irrigation at this time can minimize damage.

Trap crops release chemicals known as kairamones that are highly attractive and beneficial to insects. Kairamones produced by cucurbits include cucurbitacin, a feeding stimulant that imparts the characteristic bitter taste to cucurbits, and several floral volatiles, which are extremely attractive to adult cucumber beetles and can lure them from some distance. To be used as a lure, trap crops need to contain a much higher concentration of these chemicals than the main vegetable crop. The high concentration of kairamones in the trap crops encourages insect pests to feed on them.

Trap baits contain insect-attracting pheromones, kairamones, and other chemical attractants. Metcalf and coworkers determined that cucurbitacin can be isolated and used as a trap bait. Eugenol, primarily a pheromone attractant for corn rootworm, is being used on an experimental basis as an attractant for the spotted cucumber beetle and striped cucumber beetle.

4.3.3 Fertilizer vs. Pest

Induced resistance may be utilized for increasing the resistance temporarily as a result of some changed conditions of plants or environment, such as change in the amount of water or nutrient status of the soil (Kogan, 1994). Potassium, as a major plant nutrient can act as an inducer, especially at enhanced rates of application, of desirable biochemical alterations in the host plants to the detriment of normal feeding and development of the insects. Jeyaraman and Balasubramanian (1988) elucidated that higher dose of potash resulted in significant reduction in aphids and thrips population.

High levels of potash and phosphorus promoted the early maturity and hardening of the plant tissues, which enabled the plants to escape the injuries of the borer (Mehto and Lall, 1981). He had also observed that minimum fruit and shoot infestation of 12.60 per cent and 7.73 per cent respectively were recorded on brinjal with 100 kg of potash.

Godse and Patel (2001) studied the intensity of root knot nematode against the application of organic manures and fertilizer doses in which they concluded that brinjal root knot nematode was the minimum in organic manure treated plots.

Brinjal whitefly incidence in response to organic manures and fertilizer doses was studied by Godse and Patel (2002) and stated that there was significantly lower population in the untreated plot which was on par with FYM, neem cake, Vermicompost and half a dose of N plus normal dose of P plus double K treated plots and concluded that the lowest incidence of the borer was observed in treatment with neem cake treated plots (6.08%). However, it was on par with treatments such as vermicompost, double K_2O, FYM and half a dose of manures plus half a dose of fertilizers.

4.4 Biological Control

Biological pest control is an ideal alternative to control pests, using natural enemies such as predators, pathogens, parasites or any living organisms (biota). These are highly specific, and safe to non-target species, beneficial insects, higher animals and man. Moreover, they are inexpensive and have long term controls, cause no pollution of pest/crop agro- ecosystem. Biological control agents are available in nature and the practical methods of using them under field conditions have been investigated and exploited on a large scale in India during the last three decades.

Natural Enemies

Conservation of Natural Enemies

Weeds aid in chemical repellence, interference in movement of insects or provide alternate host for natural enemies (Altieri, 1994).

4.4.1 Biological Control in Field Crops

Rice

The spiders, Lucosa and tetragnaths spp are found to be efficient predators of brown plant hopper (BPH). Spiders help in effective management of several crop pests, especially in rice fields that are not sprayed with high doses of pesticides and devour pests. Different types of spiders and their preys are given below:

Spiders in Rice Ecoystem	Pests Controlled
Wolf spider *(Lycosa pseudoannulata)*	BPH, Green leaf hopper (GLH), White backed planthopper (WBPH) Case worm, Leaf folder and Whorl maggot
Four legged spider *(Tetragnatha spp.)*	BPH, WPH, Zigzag-leafhopper (ZLH) and White leafhopper (WLH)
Sac spider *(Clubiona japonicola)*	BPH, WPH, ZLH, WLH
Orb spider *(Araneus inustus)*	Leaf folder, BPH, WBPH and Whorl maggot

Dwarf spider *(Callitrichia formosa)*	BPH, WBPH, ZLH, Whorl maggot
Lynx spider*(Oxyopes javanus)*	BPH, GLH, WPH, Caseworm, Leaf folder and Whorl maggot
Jumping spider *(Salticid spp)*	GLH, BPH, ZLH, WBPH, Cutworm, Army worm, leafhopper Stem borer moths

Frogs and toads are also important predators of rice pests. The mirid predator, *Cyrtorhinus lividipennis*, is effective predatory bug feeding on BPH. The release of 100 mirid bugs or 50-75 eggs/Sq.m. at 10 days interval has been found to be very effective in controlling BPH. Of the. several natural enemies of rice stem borers, the egg parasitoids, *Telenomous* and *Tetrastchius*, are promising, with a high degree of natural parasitisation of egg masses under field conditions. The inundative release of *Trichogramma* at 50 thousand per ha per week is also found effective. Another indigenous parasitoid, *Platygaster oryzae*, suppresses rice gall midge under field conditions.

4.4.2 Biological Control in Other Crops

Among different indigenous species of *Trichogramma*, only *Trichogramma chilonis* Ishii (Trichogrammatidae: Hymenoptera) was predominant in India and has the distinction of being the highest produced and most utilized biocontrol agent.

Raja *et al.* (1998) noticed that the release of an egg parasitoid, *T. chilonis* at fortnightly interval significantly reduced the *L. orbonalis* damage and produced fruit yield of 20.3t per ha in brinjal crop. Sangappa, (1999) evaluated different levels of *T. chilonis* and reported that higher levels of *T. chilonis* (2.0 and 1.5 lakh/ha) were found to be effective in reducing *L. orbonalis* incidence at shoot initiation and fruit bearing stages of the *brinjal* crop.

Hanapur and Nandihalli (2003) observed maximum parasitization by *T. chilonis* against *L. orbonalis*. Inundative releases of *T. chilonis* @ 3.0 and 2.5 lakh per ha in five installments were found effective in recording less shoot infestation of 25.85 and 22.89 per cent, respectively and also recorded higher yield in potato.

A study conducted by Velmurugan and Nadarajan (2002) revealed that the predatory reduviid bug, *Rhinocoris marginatus* (Fabricius) effectively controlled the larvae of *Spoladea recurvalis* (Fabricius) on *Amaranthus* in the field cage studies.

4.4.3 Microbial Pesticides

Among the microbial insecticides, *Bacillus thuringiensis* (Bt) had received greater attention, and was widely demonstrated by the scientists all over the world. Products based on this bacterium were now the most successful of all the living insecticides.

Studies on the effect of commercial *B. thuringiensis* (Bioasp and Biolep) against fruit borers of okra by Satapathy and Panda (1997) indicated that *B. thuringiensis* formulations at 2 kg/ha were more effective and the effectiveness remained upto 15 days after treatment. Biolep was more effective when compared to bioasp. The results obtained from the work of indicated that *B t* var. *kurstaki* @ 1 ml + methomyl @ 0.80g/l of water was the best combination treatment in reducing the *brinjal fruit damage* to 3.74 per cent with an higher yield of 164.10 q/ha.

There was no adverse effect of the biopesticide, *Bacillus thuringiensis* and its products (Spicturin, Delfin, Biobit and Halt) on the growth and development of the egg parasitoids (*Trichogramma* spp.) (Srinivasan and Babu, 1998). Kumar and Sharma (2004) reported that mixture of *B t* var *kurstaki* product Dipel 8L @ 500 ml/ha could be used for the control of tomato fruit borer.

Preetha and Nadarajan, (2002) reported that *B. thuringiensis* formulation was found to be effective in reducing the fruit borer infestation on bhendi.

4.5 Botanical Pesticides

Plants were biochemists par excellence and during their long evolution had synthesized different chemicals to prevent their colonization by insects and other herbivores. These chemicals repelled the approaching insects, deterred feeding and oviposition on the plants, disrupted the behaviour and physiology of insects in various ways and even proved toxic to different developmental stages of many insects. Botanicals offered a viable alternative to synthetic insecticides as they were comparatively less toxic, biodegradable and possess diverse array of properties and were safe to non-target organisms and hence these botanicals were utilized for this study.

Rosaiah, (2001) recorded the incidence of jassids in brinjal after the application of different treatments and the results showed than custard apple and jatropha leaf extracts at 0.5 per cent concentration recorded 41.13 and 49.20 jassids / 5 plants and they evaluated the efficacy of botanicals against white fly and jatropha leaf extract at 0.5 per cent concentration recorded 42.06 adults / 5 plants followed by custard apple leaf extract (46.41). The results obtained from the findings of Chitra *et al.* (1993) revealed that petroleum ether extract of *Annona* at 0.1 per cent reduces the infestation of *Lucinodes orbonalis* by 44.8 per cent.

Kuppuswamy and Balasubramanian (1980) reported that neem kernel extract 5 per cent and neem oil 2 per cent to be effective in considerably checking the fruit borer infestation on number and weight basis of the fruits. Schmutterer, (1990) stated that neem based pesticides were safe to natural enemies due to weak contact effect in insects. The findings of Raman *et al.* (1992) revealed that Achook (a neem product) individually or alternated with insecticides was effective in controlling *L. orbonalis*.

Leaf extract and seed oil of custard apple is a traditionally well known insecticide to keep-away the pests like *Epilachna* and mealy bugs (Soni *et al.*, 2003) in brinjaThe efficacy of neem leaf extract (0.1%) was 45.51 per cent and was superior to Neem Guard 0.5 per cent and Neknool 0.1 per cent (Chitra *et al.*, 1993). Neemazal F and Neemazal T/S were found effective against brinjal shoot and fruit borer and the yield in neem products treated plots were higher than endosulfan. Neem oil 4 per cent and neem seed kernel extract 5 per cent were effective in suppression of *L. orbonalis* damage. Incidence of *L. orbonalis* was 9 to 11 per cent when neem formulations were sprayed and control plot recorded the highest damage of 43 per cent damage (Raja *et al.*, 1998).

Rajendran (1998) revealed that neem oil @ 4 per cent reduced the fecundity of eggplant spotted beetle, *Henosepilachna vigintioctopunctata* by 62.8 per cent over control. Sosamma Jacob and Sheila (1994) studied the antifeedant property of different plant products and reported that neem plant extract at 5 per cent concentration exhibited high antifeedant property against *Selepa docilis* on brinjal.

Artimesia vulgaris (0.125%) was found to be the best treatment against *Aphis gossypii* followed by, Neemall (0.25%) which also gave good control of aphid on brinjal. Syamala and Nadarajan (2006) reported that in the case of spotted leaf beetle adults, notchi followed by malaivembu were as effective as Carbaryl (0.48) with a reduction of 79.92 per cent over control.

4.5.1 Organic Pest Management

The plant products naturally occur near farm sites, and their extracts can be obtained without incurring much expense. They are safe and do not have any residual toxicity. They can be used selectively without disturbing the population of beneficial insects, parasites and predators. Further, there is a little chance of target organisms and insects developing resistance to these plant derived compounds.

Rice Pests : The following schedule may help in combating rice pests

Pest/Disease	**Organic Pesticides**
Leaf folders	Ginger, garlic and chilli extract (½ kg,1 kg and ½ kg, respectively) will give effective control
Rice grasshopper	Spray cow dung solution @ 60 lit/acre (1kg cow dung in 10 lit of water)
Yellow hairy caterpillar, horned caterpillar	*Andrographis* kashayam Neem seed kernl extract
Rice ear head bug, mealy bug	Plants with milky latex like *Calotropis, Nerium, Cacti* or *Jatropha* can be used Common plants like Neem, *Andrographis, Tinospora, Adhatoda, Vitex* and *Ocimum* may also be used
Rice black bug	Using neem cake as a top dressing helps control the pest Neem sac in water is very effective repellant
Bacterial leaf blight, brown leaf spot	Seed treatment with cow's urine Use of sweet flag extract Seed treatment with *Pseudomonas* Fumigation of seeds (Vaivadangam)
Rice blast	Spray a mixture of cow's urine and buttermilk on the affected crop

4.5.2 Botanicals for Storage Pest Control

Few of the plants (Extracts, oil, powder) that can be used for controlling, repelling and deterring the storage pests are given below in a tabular form

Organic Pesticides	Storage Pests
Neem *(Azadirachta indica)* extract over sacks. The leaves are spread in thick layers on floors of bins and along walls and crushed fruits are stuck on the walls.	Protect store food grains and seeds from Insects.
The leaves of *Trigonella foenumgreekum* (fenugreek) are used to protect the grains during rainy seasons.	stored food grains and seeds and seeds from Insects.
Artemisia absinthium is used to save the grains from the attack of the	Angoumois grain moth
Powdered neem leaves and rhizomes of turmeric	Repellant to Rice weevils.
Acorus calamus, Acorus indica, Valariana officinalis, Xanthoxylym armatum, and *Artimisia maritime* are other plant species	excellent repellent properties against the red flour beetle *(Tribolium castaneum)*
Turmeric powder, mustard oil, and common salt	Keeps free from any stored insect damage
Several medicinal plants such as *Acorus calamus, Saussrea lappa* and *Skimmia laureola*	Insect deterrents
Neem seed powder at one or two per cent mixed with wheat	Prevent the incidence of the rice weevil and Khapra beetle for over 260 days.
Use of two per cent neem oil can protect Bengal gram	For storage pests' attack
Lympholised oil from lemon, grape fruits and acid lime.	Highly toxic to pulse beetles.
Ground black pepper	Eliminate the rice weevils
Garlic oil	Wipe off the Khapra beetle
Oil extracted from the rhizomes of *Acorus calamus*	Protection from pulse and cereals pests
Wheat mixed with pounded rhizomes of *Acorus calamus,* A. *sativum,* A. *indica, Carrom capitum,* and *Xanthosylem armatum*	Keep free of pest invasion in the bins
Plant materials derived from *Cassia abscus, Ipomea hederaceae, Mentha arvenis, Mentha sylvestris* and *Ponganum harmale*	Insecticidal property against the rice weevil.

4.6 Traditional Practices for Pest Control

Some of the traditional practices followed by farmers are listed below:

- A century old-practice among citrus growers in China is to place nests of predacious ant *(Decophylla smaragdini)* in orange trees to reduce insect damage. The citrus growers even install interconnecting bamboo rods as bridges for the ants to move from tree to tree.
- Ducks, fish, frogs and snakes are traditionally used to control insects in lowland rice cultivation.
- Oil: Oil extracted from the seeds like maize and groundnut is used for treating seeds against insect attack. The oil changes the surface tension of protoplasmic contents of the developing embryo and regulates the egg density on the oil treated seed, thereby blocking the embryonic development. The following are some examples:
- Treating greemgram seeds with groundnut oil *(Bml /kg)* reduces the density of eggs of *Collosobruchus chinnensis* where 90 per cent of the eggs on the treated seeds die within 48 hours as a result of coagulation of protoplasmic contents in the embryo.
- Cowpea seeds treated with (10-15ml/kg) soybean or corn oil shows good control against C. *chinensis.* It also results in improved appearance and palatability of the seed without any change in germination.
- Cowpea seeds treated with neem kernel oil at 8ml/kg is effective in reducing oviposition of Bruchids and the seeds can be protected for 3-4 months.
- Non-edible oils like neem, castor, karanja, Undi, etc. show good repellent action against pulse beetle.
- Oils of clove, cedar wood and karanj at 25-100 ppm and acetone extracts of pepper provides protection upto 30-60 days to wheat seeds against *Sitophilus oryzae* and *Rhizopertha dominica.*
- Mixing kernel oil (0.1 per cent), kernel powder (1 per cent), cake (5 per cent), leaves (I per cent) of neem and leaves (0.5 per cent) and gum (1 per cent) of babul with wheat seeds provide protection against *Corcyra cephalonica.*

5. Conclusion

Organic pest management is useful to manage the pests in Agricultural ecosystem. Use of insecticides leads to the biomagnifications of insecticide residue in human beings. Adoption of organic pest management in crops provides pesticide free, tasteful, qualitative grains, fruits and vegetables, which will fetch international market and additional remuneration to the farmers. In addition the soil health is maintained. Though the yield in organic farming may be less in the initial periods compared to inorganic farming, considering the long range effects, it surpasses the inorganic farming and helps in preservation of the biomass, biodiversity and protection of the environment and a sustainable agriculture is possible with this method. The Vedic agriculture has already proven the beneficial effects of the good old organic farming and organic pest management.

References

Altieri, M.A. 1994. Biodiversity and Pest Management in Agroecosystems. Haworth Press, New York, USA, pp. 185.

AVRDC, 1984. Sprinkler irrigation for diamondback moth control. AVRDC progress report, pp. 27-28.

Chitra, K.C, S. Janardhan Rao, P. Kameswara Rao and K. Nagaiah. 1993. Field evaluation of certain plant products in the control of brinjal pest complex. *Indian. J. Ent.*, 55(3): 237-240.

Godse, S.K. and C.B. Patel. 2001. Effect of organic manure & fertilizers doses in management of root knot nematode *(Meloidogyne spp.)* of brinjal. *Plant Prot. Bull.*, 53(1+2): 46-49.

Godse, S.K. and C.B. Patel. 2002. Effect of organic manures & fertilizer doses on the incidence of brinjal shoot and fruit borer (*Leucinodes orbonalis* Guenee) *Pestology*, 27(5): 9-12.

Jeyaraman, S and R. Balasubramanian. 1988. Role of potassium treatment on yield and incidence of pests and diseases in chilli. *J. Pot. Res.*, 4(2): 67-70.

Kogan, M. 1994. Plant resistance in pest management. In: R.L. Metcalf and W.H. Luckmann (Eds). Introduction to IPM. John Wiley and Sons. New York, USA, pp. 73-127.

Kumar,J. and S.D. Sharma. 2004. Efficacy and economics of *Bacillus thuringiensis* var kurstaki for management of Helicoverpa armigera on tomato (*Leucinodes orbonalis*) in lower Kulu valley,

Kuppuswamy, S. and M.Balasubramanian, 1980. Efficacy of synthetic Pyrethroids against brinjal fruit borer, *Leucinodes orbonalis* Guen. *South Indian Hort.*, 28: 91-93.

Preetha, G. and L.Nadarajan. 2002. Evaluation of IPM modules to manage the pests of bhendi in Karaikal regions. M.Sc. thesis Submitted to the Department of Agricultural Entomology, PAJANCOA and RI, Karaikal. Puducherry.

Raja, J., B. Rajendran and C.M. Pappaiah. 1998. Management of egg plant S&FB, *Leucinodes orbonalis* Guenee. Proc. *I. Nat. Symp. Pest Mgmnt. Hort. Crops*, IIHR, Bangalore, 15-17. October, 1997. pp. 84-86.

Rajendran, B.1998. Effect of neem oil and endosulfan on the oviposition and hatchability of *Henosepilachna vigintioctopunctata* (Fab.) *Proc. I.Nat. Symp. Pest Mgt. Hort. Crops*, Bangalore, pp: 106-107.

Raman, K., S. Ganesan and B.N. Vyas. 1992. Studies on the effect of neem rich insecticides on feeding, development and control of cotton and vegetable posts. Neem Newsletter, 9(4): 56-57.

Rosaiah, B. 2001. Evaluation of different botanicals against the pest complex of brinjal. Pestology, 25(4): 14-16.

Sangappa, N. 1990. Management of brinjal shoot and fruit borer, *Leucinodes orbonalis* Guenee. M.Sc., (Ag.). Thesis, University of Agricultural Sciences, Dharwad.

Satapathy, C.R. and S.K. Panda. 1997. Effect of commercial *Bt* formulations against fruit borers of okra. Insect Environment, 3(2): 54.

Schmutterer, H., 1990. Side effects of neem products on insect pathogens and natural enemies of spider mites and insects. *J. Appl. Entomol.*, 121(2):128.

Soni, A.K, G.V. Raman, V. Daniel, G.L. David Krupadanam and G. Sriman narayana. 2003. Isolation of active principles from *Annona squamosa* and Pongamia *Pinnata,* Insect-Antifeedant activity and field studies on some Agricultural crops. *Proc-Nat.Sem.Cont. Insect plant insecticide*, 5:47-54.

Sosamma Jacob and M.K. Sheila 1994. Studies on the antifeedant activity of some plant products against the leaf caterpillar, *Selepa docilis* Butl. on brinjal and wooly bear, Pericallia ricini on castor. *Indian J. Ent.*, 56(3): 276-279.

Srinivasan, G. and P.C.S. Babu. 1998. Life table studies on shoot and fruit borer, *Leucinodes orbonalis* Guenee (Lepidoptera: Pyralidae) on Brinjal, tomato and potato. *Proc. I Nat. Symp. Pest Mgt. Hort. Crops,* Bangalore, pp. 94-97.

Syamala, M and L. Nadarajan. 2006. Bioefficacy of botanicals against brinjal spotted leaf beetle, Henoepilachna vigintioctopuntata Fabricius (Coleoptera: coccinellidae) in Karaikal. M.Sc.,(Ag.) thesis. Pp.135.

Velmurugan, R. and L.Nadarajan, 2002. Seasonal abundance of lepidopteran pests in amaranth and their management with the predator *Rhynocoris marginatus* (Fabricius) (*Hemiptera: Reduviidae*). M.Sc. (Ag) thesis. PAJANCOA.

❏❏❏

6

Integrated Pest Management in Tea in South India : Problems and Prospects

BABU, A.
UPASI Tea Research Foundation, Nirar Dam B.P.O. , Valparai-642 127

1. Introduction

Tea is grown in the humid belt of the Western Ghats with varied agro-climatic conditions in south India. The weather conditions in the different planting districts vary distinctly. Majority of the tea growing areas in the North India are in flat land while in south India, it is cultivated in hill slopes at an elevation ranging between 500m to 2400m. Tea, as a perennial crop is known to be infested by a number of insect and mite pests. Most of the tea pests are highly seasonal; many attack tea only during dry season while a few are abundant in wet weather; there are a few perennial pests too. The first and foremost aspect for tackling the pests is to have a better understanding about the pest itself. If this aspect is known we can proceed scientifically in controlling the pest attack without leading to residue problems in made tea

The major pests *viz.* thrips, tea mosquito bug, red spider mite and shot hole borer cause widespread damage, resulting in appreciable crop loss apart from increasing the cost on pest control. Data are available on the bio-ecology and crop loss caused by various pests such as red spider mites, thrips, tea mosquito and shot hole borer. They are responsible for an average crop loss of 12 - 15 per cent. Pest management in tea is mainly achieved by the use of synthetic chemicals.

To overcome the current situation, it is necessary to look for alternate strategies by adopting non - chemical control measures. Indiscriminate use of pesticides and improper execution of pest control measures will lead to undesirable problems in tea gardens. At this juncture, switching over to non-chemical control strategies on large scale is essential in order to compete in international market for fetching good price for our products. Pesticides aimed at destructive insect populations have the potential to alter the balance of the tea ecosystem by destroying the natural enemies (non target organisms) such as predators and parasitoids. Continuous use of synthetic chemicals for the control may also lead to development of resistance in targeted organisms. The enormous crop loss caused by pests requires implementation of efficient, eco-friendly control strategies. Biopesticides are considered as environmentally safe, selective, biodegradable economical and renewable alternatives for use in IPM programme especially under organic farming systems.

During the last few years research work has been carried out and the data has been generated on different biological control methods including microbial control (use of fungus, bacteria), semiochemicals (pheromones and kairomones) and other biologicals. Even though, neem based products are used in many crops, their use is very much limited in tea, except for certain Azadirachtin based commercial formulations. The efficacy of Neem Kernel Aqueous Extract (NKAE) has been evaluated against important tea pests both in lab and field conditions to determine the effective concentration for controlling them. Although different pest control methods are adopted over the past several decades, there is a need for a closer look at the current status of important pests, to arrive at an integrated pest management programme

Development of mass multiplication technology of the entomopathogens (microbes), natural enemies (major predators) and use of biologicals will be definitely helpful in reducing the usage of synthetic insecticides for the control of tea pests. An attempt has been made in this paper to provide a through knowledge about the important tea pests, seasonal abundance, damage potential and integration of different biological control strategies as a package for implementing them in tea gardens.

2. Tea Thrips *Scirtothrips bispinosus* Bagnall

Scirtothrips bispinosus (Bagnall) is an endemic species of peninsular India, causing severe damage to tea plantations especially in Vandiperiyar-Peermade area of Idukki District in Kerala. It has been reported to attack tea as well as coffee Tea fields recovering from pruning is more prone to the attack by this thrips. Leaf surface becomes uneven, curly and matty, exhibiting parallel lines of feeding marks on either side of the midrib. Heavily infested fields sometimes acquire a bronze color. Damaged terminal shoot may be discolored, stunted, and deformed. The threshold density is determined as three thrips per shoot, above which they cause economic damage leading to considerable crop loss.

The life cycle consists of an egg stage (incubation period of 3-4 days), two larval instars (larva I - 3 days, larva II- 5 days), a short pre pupal stage (< 1 day), a pupa (3 days). The larvae resemble the adults in general body form though they lack wings and are smaller. Larvae feed in groups, particularly along the leaf midrib and veins, and usually on young leaves and buds. Larval development time is determined principally by the suitability of temperature. Total duration of life cycle is about 14- 20 days depending on the weather conditions.

2.1 Symptoms of Thrips Infestation

Lacerations of tissue and appear as streaks, leaf surface becomes uneven, curled and matty, appearance of parallel lines on either side of mid rib is due to feeding marks in buds which becomes prominently visible when the leaf unfolds. Yellowing of leaf margins is the typical symptom of thrips infestation.

2.2 Damage

More damage in tea recovering from pruning and delayed tipping causes appreciable crop loss and also affects the quality of made tea.

2.3 Monitoring

Monitoring of thrips has to be done by inspecting the pluckable shoots (three leaves and a bud stage) and counting the number of larvae and adults separately. Shoots should be collected from the side branches, plucking table and below the plucking table. The shoots should be collected at random from different bushes plants in the field. Adults can be monitored by the use of yellow sticky traps even at the initial stage, thus could be prevented from the sudden flare-up by adopting control strategies.

2.4 Economic Threshold Level (ETL)

There are several recommendations for the economic threshold for different crops. One should remember that the economic threshold should be validated for each agro climatic conditions of different regions based on the thrips incidence, the anticipated crop loss, previous year's incidence and environmental conditions. Based on the research, the threshold is fixed as 3 thrips per shoot. Appropriate control measures have to be adopted to reduce economic loss due to thrips infestation if thrips population is around 3 thrips per shoot.

The monitoring of thrips population should be initiated from September onwards when they start their appearance in a small scale. The pest population should be monitored even after the preliminary rounds of spraying and the affected section should be marked to facilitate spot spraying instead of blanket spray, which is not necessary at that particular time.

2.5 Control Strategies

2.5.1 Chemical Control

Timely adoption of recommended chemicals having label clearance from the CIB is essential to contain the population of thrips. The follow-up round should be given at 7-10 days interval so as to have the pest under check. Care should be taken to avoid spraying chemicals either before plucking or without giving the stipulated safe harvest interval which may lead to taint/reside problems in made tea.

2.5.2 Pesticide Recommendations

In view of pesticide residue problems general awareness of pesticide residues around the globe, it is very important to use pesticides as little as possible by integrating different control strategies by adopting integrated pest management (IPM) programs. Only the recommended pesticides with clearance from CIB have to be used in rotation at stipulated intervals. The selection of pesticides depending on the nature of the pest attack is very critical in achieving satisfactory control of pest.

The success of chemical control of thrips in tea is primarily depending on application method. It is necessary that the pesticide is reaches the target. Hence spraying has to be done in such a way that it covers both sides of the tea leaves and side shoots, where the majority of the thrips are located.

2.5.3 Biological Control

There are several natural enemies that help in the control of thrips. However, none of them alone can reduce the thrips populations to a low, non-economical density. Furthermore, the intensive use of pesticides in the tea affected adversely on the ecosystem natural enemy's activity. There is a need for more study on the role of natural enemies in the tea ecosystem and to enhance natural enemy populations by planting a natural barrow row of flowering plants along the borders/vacant places. The entomopathogenic fungus, *Verticillium lecanii* was also effective in controlling the thrips. Yellow sticky trap was found useful in monitoring the thrips population even at an early stage of infestation. It could be incorporated into the IPM schedule and the use of pesticides should be need based, and with sufficient safe-harvest intervals in order to keep residue levels of pesticide below stipulated maximum residue limit.

3. Red Spider Mite *Oligonychus coffeae* (Nietner)

The tea red spider mite, *Oligonychus coffeae*, which was considered as a "dry weather pest" until the last four to five years in south India, has gained importance in the last few years due to its sudden outbreak in the tea growing areas of south India (Muraleedharan *et. al.* 2005). It is the largest of all the mites and can be easily seen by the naked eye. They normally infest the upper surface of mature leaves, though in extreme cases they may invade the lower surface as well as young leaves and consequent to their feeding, the maintenance foliage turns ruddy bronze, making the infested fields distinct even from a distance. Adopting control measures in proper time will prevent the total defoliation of bushes. At present it assumed the status of major pest causing severe damage to the tea plantations. Most of the tea estates are forced to spend a huge amount for the control of red spider mite.

The eggs are laid along the veins of leaves during the growing season. Red spider mite eggs are round and extremely large in proportion to the size of the adults. The maximum activity of spider mite peaks during the warmer months and can develop rapidly during this time, becoming full-grown in as little as a week after eggs hatch. After hatching, the egg shells remaining on the leaves, helps in diagnosing spider mite infestation. The fast development rate and high egg production can lead to extremely rapid increase in mite population. They feed more under dry conditions, as the lower humidity allows them to evaporate excess water they excrete. At the same time, most of their natural enemies require more humid conditions and are stressed by arid conditions. Furthermore, plants stressed by drought can produce changes in their chemistry that make them more nutritious to mites.

3.1 Assessment of Population of Red Spider Mites

Sampling for red spider mite consists of collecting leaves and counting number of mite eggs, nymphs and adults. The current ecomomic threshold Level (ETL) for red spider mite is 4 mites per leaf. Large-scale sampling should be done by demarking 1 ha area into four or six plots and taking 50 leaves randomly from each plot and count the number of egg, nymphs and adults. Care should be taken to take into account even the presence of a single mite in a leaf as infested. If 30% of the leaves contain mites immediate spraying should be taken up.

3.2 Control Measures

Lime sulphur is a potent acaricide useful in controlling red spider mite population, provided it is prepared in such a way that the Polysulphide-S content should be above 10%. It is rather inexpensive and it can be integrated with other chemical schedule. However, continuous usage of Lime Sulphur is discouraged because of its poor ovicidal activity and tainting nature.

The new herbal product Exodus derived from leguminous plant *Sophora favescens* was found effective against red spider mite. Mineral spray oils manufactured from Paraffin base has also found to be effective and is found safer to biological control agents also. New acaricides like Propargite 57EC, Fenpyroximate 5 %EC are found to be more effective and their efficacy lasted for more than 3 weeks. However, we have to go for an integrated approach by incorporating the various eco-friendly, non-chemical methods available, to achieve efficient control of this pest.

The entomopathogenic fungus, *Paecilomyces fumosoreus* has been isolated from the flushworm infesting tea and it has been formulated into a wettable powder (UPASI TRF Strain) and tested against the red spider mite It was found effective in controlling the red spider mite. The fungal formulation is now commercially available under the trade name "Mycomite". The formulation has to be applied@ 2000g/ha in 450 litres of water, preferably using a hand operated/high volume sprayer. The spray solution should be filtered through muslin cloth to remove extraneous materials and to prevent clogging of the nozzles. Application of "Mycomite" has to be done early in the mite season when the red spider mite incidence is low. The ideal time for spraying will be early morning or late evening when the relative humidity is around 70-80 %. All the life stages of the red spider mite are susceptible to the fungal pathogen and mortality could be noticed after 5-6 days of application. However, close monitoring of mite population may be necessary even after "Mycomite" spraying and spot application of Lime sulphur is recommended to achieve better control. It must be ensured to get minimum period of 10-12 days between fungicide application for the control of diseases and Mycomite spraying.

Any delay in taking control measures lead to further build-up of the pest, which results in serious damage to tea bushes. The spray volume recommended is 350-400L/ha (for power sprayers) 450-500L/ha (for knapsack sprayers) and 600L/ha (for Honda sprayers). Thorough drenching of maintenance foliage is essential for efficient control of mites, because large number of eggs, nymphs and adult mites are seen below the plucking table. Addition of no-ionic wetting agent @ 5ml/10L is recommended for better control during wet weather. As in the case of any crop pests, integrated control measures are required to reduce the attack. The small-scattered pockets from where the infestation starts must be identified. Efforts should be made to keep these areas marked and to avoid spreading to other areas.

4. Tea Mosquito Bug, *Helopeltis theivora*

The Tea Mosquito Bug *Helopeltis theivora* is yet another important pest causing serious threat to the tea plantations especially in Central Travancore region. They prefer dark, humid and shaded areas. They are very much active during the early hours of the day and late evenings. They hide themselves among the weeds and weed hosts during the day time. Incidence will be more during July - December with a peak during September/October. Initial infestation starts in the fields near jungle and swamps. The chinary and hybrid jats are more susceptible. Eggs are laid in tender stems, petioles. One can identify the egg laid spots by close observation of the two unequal respiratory horns of the eggs, exposed from the stem or petiole. The incubation period of the egg is about 5-7 days. It passes through 5 nymphal instars within a period of 9-12 days depending on the favorable weather conditions. Adults resemble mosquitoes and are black colour. Adults and nymphs suck the sap from the buds, young leaves and tender stems by puncturing with needle like rostrum. Intensive feeding causes curling of leaves, badly deform, remain small and shoot dry up and crop loss is near total. Life cycle is completed within 19-29 days.

4.1 Assessment of Infestation

Percentage of infestation has to be assessed by collecting 100 shoots (three leaves and a bud) at random from the plucking gang of different fields and counting the infested shoots from the hundred shoots. Even a single new puncture is counted as infested. Old punctures should not be included as infested. If the average infestation is above 5 % immediate spraying should be done soon after plucking with a follow-up round at 7-9 days interval.

4.2 Control Measures

Tea mosquito bug incidence may be monitored regularly and suitable control measures should be taken up without any delay. Pesticide application using Quinalphos @750 ml/ ha may be taken up for the control of *Helopeltis*. Follow-up rounds may be done after 7-10 days in the severely affected fields.

Cultural Control

Keeping the fields free from weeds is the most important parameter to minimize the TMB infestation. Removing the stubs intensively while plucking will remove eggs and thus we can reduce the next generation. Black plucking of affected bud can also be done to reduce the egg population.

5. Shot-Hole Borer *Euwallacea fornicatus* (Eichhoff)

Shot-hole borer, *Euwallacea fornicatus* is one of the important pest of tea in the middle and low elevation tea growing districts. The females construct galleries in tea stems, leading to branch breakage and consequent indirect crop loss. They form the galleries in the primary branches formed after pruning. However, the attack may also see on the old wood, even at the collar region. Pruning helps in the removal of the affected stems and adult beetles. The first year fields will be free of SHB infestation due the physiological changes taking place in the bushes immediately after pruning. We can notice the new infestation only from the end of the 2nd year. Being a perennial pest, it could be managed only by adopting system of pest management.

5.1 Assessment of Population

The infested fields have to be divided into 2ha blocks and from each block, collect 200 cuttings of pencil thickness (1.5cm diameter and 20 cm long) at random. The percentage of attack in each block has to be pooled and take the average infestation of all blocks. Control measures have to be taken up if the average infestation is above 15 percent.

5.2 Control Measures

Integrated Pest management is the best way. It is better to avoid pruning during the wet weather. Removal of badly affected branches while pruning is another way of preventing the migration of adult beetles from one filed to the other. Chemicals like Quinalphos 25EC @ 750 ml + Dichlorvos 76 EC @ 250ml/ha have to be sprayed on the pruned frames and prunings. Two applications are required during April and December for efficient management of this pest.

5.2.1 Mid Cycle Control

Assess the SHB infestation at the end of 2nd year after pruning, and it would be advantageous to commence mid cycle application in 3rd & 4th year fields, if the average infestation is above 15%.

5.2.2 Biological Control

Two rounds of the entomopathogen, *Beauveria bassiana* (Biopower) @ 1500g/ha is recommnede for spraying during end May and October, when the humidity levels are high. Studies conducted at UPASI TRF revealed that partially dried cut stems of a jungle plant *Montanova bipinnatifida* .an exotic plant species introduced from Mexico, attract shot hole borer beetles in the field. Placing partially dried cut stems of *Montanova* has helped in trapping the beetles and reducing the incidence of this pest. Such dried stems could be used in SHB infested 3rd and 4th year fields during October to April to attract the adult beetles. Stems of 3-4 weeks old 90cm long and 25-30cm thickness may be used. Place 4 cut stems in $100m^2$ and an average of 400 cut stems are required per ha. Replace the stems once in three weeks.

6. Conclusion

Even though, a number of synthetic chemicals are available for pest control in tea, the choice is limited for widespread application owing to the Maximum Residue Limits (MRLs) declared by European Union (EU) and Codex Alimentarius Commission and EPA of USA. Hence need based, judicious and safe application of pesticides have become the vital aspect of chemical control under the integrated pest management strategy, which involves development of IPM skills. Monitoring the crop health, observing the ETL and identification and conservation of biological control agents including entomopathogens, evaluation of their field potential, before taking a decision on the sue of chemical pesticides as last resort. The success of any control measure under IPM context depends on the various parameters such as (i) use of pesticides only when it is absolutely necessary, (ii) use of recommended chemical with label claim on tea, (iii) maintaining correct dosage and dilution rates, (iv) giving adequate pre-harvest interval, (v) change of pesticides in the subsequent rounds, (vi) incorporation of biologicals including entomopathogens.

Adoption of such non-chemical control methods to evolve an integrated management programme against pests infesting tea will form an integral part of sustainable tea cultivation.

❑❑❑

7

Pest Problems and their Eco-Friendly Management Techniques in Cardamom

KUMARESAN, D.
Former Senior Scientist, Indian Cardamom Research Institute, Spices Board, Mayiladumbara.

Cardamom is depredated by an array of pests from the seedling stage in the nursery to the yield stage of main plantation. The pest problem has become a major factor limiting the crop productivity. Nearly 60 species of insect and non- insect pests infest on cardamom (Gopakumar and Chandrasekar, 2002). One of the major constr aints in the production of cardamom is the excessive damage caused by pests; the major ones being *Sciothrips cardamomi* (Ramk) and *Conogethes punctiferalis* (Gurnee). Based on severity of infestation, pests are categorized as major, sporadic and minor pests the former includes thrips; shoot panicle and capsule borer (SPCB) root grub and white fly, hairy caterpillars, mites and nematodes recorded as sporadic pests. The minor pests include rhizome weevil, shootfly, cutworm, aphid, scale insects and leaf folders.

1. Cardamom Thrips: *Sciothrips cardamomi* (RAMK)

In the order of importance with respect to the extent of damage caused, the Cardamom thrips ranks first among the insects pests of

Cardamom in India whereas in countries like Srilanka and Guatemala this is not a serious pest.

This insect was reported in India during the 1934 – 35 crop seasons in Anamalai hills, Coimbatore Districts, Tamilnadu as being responsible for the extensive scabbing of capsules and identified as *Taeniothrips cardamomi* RAMK. (Ayyar, 1935; Ayyar and Kailasam, 1935) and later in 1936, the authors described the damage caused by insect. In the succeeding years this pest was noticed throughout the cardamom tracts.

1.1 Nature of Damage

The larvae and adults lacerate tissues from leaf sheaths, panicles, flowers, tender capsules and suck the exuding sap resulting in qualitative and quantitative loss. Infestations on panicle results stunted growth, and shedding of flowers. Laceration of tissues from tender capsules leads to the formation of small scabs, which develop as prominent callus growth on the capsules such capsules appear as stunts, malformed and shriveled with gapping slits on the skin. This is popularly called as "Cardamom itch" (Ananthakrishnan, 1984; Nair, 1978).seeds of infested capsules are poorly developed and do not germinate and lack the usual aroma and fetch low price in the market. The content of 1, 8 cineole was reported as higher in the oil extracted from thrips damaged capsules (> 50% damage) contributing to the undesirable camphoraceous flavor and terpenyl acetate responsible for sweet, spicy flavor was at low concentration in the oil extracted from thrips damaged capsule as compared to healthy. The number of capsules per unit weight was more due to thrips damage as compared to healthy capsules and avoidable crop loss is worked out about 45 to 48 percent.

2. Shoot, Panicle, Capsule Borer (SPCB)
(*Conogethes punctiferalis* Guen.)

SPCB is a maj or pest in nurseries and main fields in all the cardamom growing tracts in Western Ghats. Larvae of this pest bore into panicles/capsules/shoots and grow by feeding on the internal tissues. Infested tillers and panicles dry off and capsules become empty. The adult is a medium sized moth with orange yellow wings having several black dots over it. A female moth lays about 20 – 35 eggs singly or groups of two or three on leaf margins or dry leaf sheath or along leaf veins. (Gopakumar and Chandrasekar, 2002). Full grown larvae are pale pink, 30 - 35 mm long and crawl to a place near to the bore hole within the shoot. Appearance of extrude frass material at the bore hole confirms the presence of larva with in the shoot. During the ensuing pre- pupal period of a day or two, they remain quiescent with in the self - made cocoon and soon become pupae. Pupa is 10 - 15mm long, pale green soon after its formation and gradually changes to deep brown. Pre-pupa and pupa are non feeding, harmless stages. Pupal period lasts 10 - 12 days. Its life cycle during March - October is completed within 41 - 68 days, while during the winter months of November - February, it prolongs to 123 days (Varadarasan *et al.*, 1990), Ravichandran (1988) reported that the life cycle of SPCB during February -April is completed 38 - 44 days at Thadiyankudisai, Tamilnadu.

Castor was the preferred host of *C. punctiferalis* causing very serious damage up to 16.4 to 82.6 per cent but on cardamom it was only 10 per cent of yield loss *C. punctiferalis* preferred castor rather than cardamom. The larvae settled on castor were 20.40 per cent whereas it was 13.60 per cent on cardamom.

3. Root Grubs: *Basilepta fulvicorne* Jacoby

Root grub is a subterranean pest of cardamom. It damages roots and thereby obstructs uptake of nutrients and also as a pre- disposing factor for root pathogen leading to yellowing of leaves. (Gopakumar *et al.*, 1987). The pest is recorded both in nurseries and plantations of all cardamom growing tracts in Western Ghats. However this pest causes more damage to seedlings in the nursery in Karnataka cardamom growing tracts.

4. Cardamom Whitefly – *Kanakarajiella cardamomi* (David and Sundararaj)

The species of whitefly reported to infest cardamom are Dialeurodes cardamomi David and Subr. (The genus is renamed as *Kanakaraj* David and Sundararaj, 1993), *Aleuroclava cardamomi* (David and Sub.), *Aleurocanthus* sp; *Bemisia tabaci* (Genn) and *Cockerella diascoreae* Sundararaj and David (Selvakumaran and Kumaresan, 1993) however *K. cardamomi* is destructive to cardamom.

Hot humid climate is favorable for its multiplication. With the onset of monsoon, the insect gets automatically reduced in number.

5. Nematodes

Plant parasitic nematodes belonging to 28 species and 19 genera are reported.

Among them, root knot nematodes (*Meloidogyne spp*) pose a serious problem in nurseries and also in plantations.

5.1 Nature of Damage

- *Primary nursery*

  Seedlings infested with root knot nematodes exhibit stunting, poor tillering, yellowing and drying of leaf tips and margins along with heavy galling of roots.

- *Secondary nursery*

  Seedlings harbouring root knot nematodes show rosette appearance and irregular arrangement of leaves. In heavily infested seedlings curling of unopened leaf which failed to emerge and form a whip like structure stunting along with yellowing of leaf margins and 'tip burn' are common aerial symptoms.(Eapen, 1994)

- *Plantation*

  In general, nematode infestation is not evenly spread over a field. It occurs in patches and gradually spreads. The affected plants appear stunted in growth and their leaves are small and rolled up (tunneling of leaves). Infested plants show excessive branching near the root tips or along the entire root at different level.

- *Crop loss*

  Eapen (1994) reported a crop loss of 46.6 percent under stimulated micro plot studies with a population load of 4 nematodes / 100 cm^3 soil. (*Meloidogyne* spp*)* *Pratylenchus spp* is commonly seen in cardamom - coffee mixed plantations while *Radopholus similis is* prevalent in cardamom - arecanut mixed gardens.

  Meloidogyne incognita is the predisposing factor *Rhizoctonia solani* infection, causing damping off and rhizome rots, prevalent in cardamom nurseries and plantations.

  Cardamom plants infected with 'katte' mosaic virus supported 5- 10 times more *M.incognita* populations in cardamom tracts, Karnataka. (Ali and Venugopal, 1992).

6. Pest Management Practices

Pests are one of the important biological components in cardamom ecosystem. A pest which causes major damage every season, unless controlled, is called 'major'/ key pest while those causing damage infrequently are the occasional/ sporadic pests. In general, IPM practices to be followed mainly for key pests, initially. Crop loss assessment is the main criteria to decide the 'key' pest status in any crop ecosystem. Among the key pest in cardamom ecosystem, Thrips and SPCB contribute major crop loss; root grubs and whitefly are occasional pests. Among non insect pests, nematodes recorded as sporadic pests but it is a major pest in cardamom nurseries.

Various components in cardamom pest management are outlined.

6.1 Cultural Control

Cultural control practices are considered important to suppress pest populations and damage in the integrated pest management programmes. Cultural control may not themselves reduce pest population below acceptable levels but may be of benefit in reducing losses due to pests through the magnitude or frequency of outbreaks or manipulating habitat.

7. Thrips

Cardamom thrips need enclosed niches for their shelter, food and reproduction. The dried leaf sheaths and panicles provide convenient shelter for thrips during summer resulting in higher population build-up. The removal of these dried leaf sheaths, panicles in the plant during summer constitute pruning in cardamom, because of pruning the niches available for their higher population build up could be dealt with minimum insecticide load into cardamom ecosystem.

The pruning operation recorded a reduction of thrips infestation to an extent of 36 percent damage on capsules over unpruned plots. Planters' adopt a general trashing (including pruning) operation; otherwise it is called as field sanitation, after the receipt of south- west monsoon showers during June. Hence the build up of thrips population in lower leaf sheaths and bracts in the panicles providing convenient shelter for feeding and migration goes unabated till trashing is done at a very late stage. Hence, the pruning is advanced during the summer, as a cultural operation could reduce thrips population.

The cultural operation followed with insecticide application could serve as an economical input and reduce the insecticide load in cardamom ecosystem.

7.1 Spacing

Closer spacing between plants is found to increase thrips infestation and SPCB infestation.

7.2 Shade Regulation

Intensity of shade plays a predominant role in determine the population density of major pests in cardamom ecosystem; Cardamom grows luxuriously in cool, shady environment. The shade canopy provides suitable condition by maintaining the humidity and evaporation at required level. To get a sustainable yield of cardamom, shade has to be maintained at 40 – 60 percent level according to the lie of land and type of irrigation.

A good shade pattern is essential for high productivity in cardamom. However, the low and high shade pattern in cardamom ecosystem plays a major role in pest density.

Low shade pattern creates a congenial condition for multiplication of thrips population and also shoot borer population. Root grub beetles recorded more on low shade area which results in higher damage certain trees like *Macarnga indica Cleodendron infortunatum* are alternate hosts for several caterpillar pests. Dadap trees, *Erythrina indica* Lam and *E. lithosperma* Mig. are hosts for root knot nematodes (*Meloidogyne* spp) and root grubs (*B.fulvicorne*).

7.3 Trap Crop Strategy

Trap cropping is the use of an attractive crop to lure pest where, the pest can be controlled thereby minimizing the chemical sprays on the main crop. The laboratory experiments on the host preference of *C. punctiferalis* revealed castor as the most preferred host with a preference of 1.5 times more than cardamom. Such an observation was earlier reported as C. *punctiferalis* affects castor heavily (16.4 to 82.6 %). The reproductive rate of *C. punctiferalis* is about 20.9 times in castor with a life cycle of 29 days as against 6.48 times with 38 days in cardamom C. *punctiferalis* found to prefer 2.1 times more than the cardamom and hence the castor can very well be exploited as a trap crop in cardamom.

7.4 Field Sanitation

Growth of weeds and wild host plants in and around cardamom ecosystem should be controlled and this will help to reduce the major pests in cardamom.

7.5 Thrips Management

Avoid the following plants growth in the vicinity of cardamom ecosystem. *Panicum longipes; Hedychium flavescens; H.coronarium; Ammomum cannaecarpum; A. involucratum; Colocasia antiauorum; Zingeber* sp; *Curcuma pseudamontana; Costus speciosus; Cririum* sp; *Globba ophioglossa* and *Alpinia galangal.*

8. SPCB Management

The pest is polyphagous and the following plants recorded as alternate host and these host plants may be avoided in the vicinity of cardamom plantation.

Guava (*Psidium guajava* L); Mango (*Mangifera indica* L); Peaches (*Prunus persica* Benth and Hook); Pomegranate (*Punica granatum* L); Jack (*Artocarpus integrifolia* L); Ginger (*Zingiber officinale* Rosc); Turmeric (*Curcuma longa* L); Avocado (*Persia gratissima* Gaettn); Mulberry (*Morus alba* L); Loquat (*Eribotrya japonica* Lind); Pear (*Pyrus communis* L); Sorghum (*Sorghum bicolor* L); Cocoa (*Theobroma cacao* L); Castor (*Ricinus communis* L); Tamarind (*Tamarindus indicus* L);Soap nut (*Sapindus emarginatus* Vahl.); *Anona cherimola.* Nambiar *et al.*, observed *Curcuma aromatica, C. amada, Alpinia* sp., *Amomum* sp., *A. subulatum, Hedychium coronarium, H. flavescens* and *Alpinia galangal.*

9. Root Grub

Beetles are polyphagous; Jack (Artlocarpus *heterophylla*), Indian almond (*Terminalia catopa*) Mango (*Mangifera indica*), Guava (*Psidium gujava*) Ficus (*Ficus indica; F. bengalensis*) Cocoa (*Theobroma cacao*), Dadaps (*Erythrina Lithosperma*) etc; are certain recorded alternate hosts of the beetle.

9.1 Host Plant Resistance

The cultivars of cardamom having different growth habits of panicle pattern, vary in their susceptibility to resistance or tolerance to pests. Cv.Malabar having prostate type of panicle growth, record less thrips population and also thrips damage on capsules as compared to Cv. Vazhuka (semi - erect) and Cv.Mysore (erect). Accession no. 32 (Maintained at ICRI, Myladumpara) recorded less thrips damage (32%) on capsule Singh *et al.* (1995) reported the clone number SKP- 97 (Cv.Malabar1) and local Cv.Malabar were moderately resistant to thrips.

9.2 Physical Control

9.2.1 Manual Method

Population of foliage feeding hairy caterpillars can be reduced to a great extent by manual removal of larvae and pupae. Root grub beetles at the period of their massive emergence, (March - April and August - October), alighting on cardamom plants, can be easily collected using an insect net and destroyed. A labour could be collected 2500 - 3000 beetles a day. Such massive collection and destruction of beetles drastically reduces the grub population in to a low level.

9.3 Soil Solarisation

Soil solarisation during summer is found quite effective in reducing the *Meloidogyne* spp. population in the infested soil and also in soil media for cardamom nursery. Raking up the soil around the cardamom plants and also in between the row of the plants and subsequent application of insecticides during May - June is an effective method for root grub managements (Gopakumar *et al.* 1987).

9.4 Light Traps

Fluorescent light traps are useful in attracting the moths of hairy caterpillars and adults of SPCB. These traps are useful for monitoring the activity of the pests and also as a tool in their suppression.

9.5 Yellow Sticky Trap

Adults of whitefly are very much attracted towards yellow colour and this behaviour is well exploited for trapping them on a yellow surface coated with sticky material. Yellow sticky traps made of metal sheets painted yellow and coated with castor oil have been recommended to trap and kill adults. Yellow fluorescent polythene sheet coated with castor oil / white grease is fixed in the shade trees of cardamom plantation, to a height of cardamom plant facing towards eastern side, this may serve to know the population density of whitefly adults in the particular cardamom eco - system in order to plan the management practices. This practice is regularly followed by the cardamom planters.

Pest	Natural enemies
SPCB (larva)	**Parasitoids**
	An unidentified Braconid
	Phanerotoma hendecaseilla cam
	Apanteles sp
	Bracon brevicornis Wasmael
	Bracon (Microbracon) *hebator* say
	Myosoma sp
	Xanthopimpla sp
	Dolichorus sp
	Xanthopimpla australis Kr
	Friona sp
	Agrypon sp
	Apectnis capulifera Kriech
	Scambus persimilis Ashmead
	Brachymeria obscurata Walker
	Brachymeria euploeae West
	Palexorista parachrysops Bezzi
	Paligosteroptera sp
	Atherigona sp
	Anatrichus pygmaeus Loew
Pupa	*Brachymeria* sp. nr. *Nephantidis* Ghan
	B. obscurata Walker
	Agrypon sp
	Predators
Larva	*Anisolabis stali* Dourn
	Philodicus sp

9.6 Bio Control

Among the above Parasitoids, natural parasitization by *Agrypon* sp on larvae of *C.punctiferalis* was maximum (19.8 percent) during November in cardamom growing tracts of Udumpanchola, Kerala.

10. Whitefly

Under natural conditions the whitefly has been found susceptible to a number of natural enemies such as predators like *Mallada bonninensis*, unidentified neuropteran, dipteran and predatory mites; Parasitoids such as *Encarsia septentrionalis* and *E. dialeurodes*. Potential of these natural enemies as biocontrol agents of whiteflies is to be explored.

11. Thrips

Anthocorid bugs and chrysopa are reported as natural enemies on cardamom thrips in the field. However, its potential is yet to be explored.

11.1 Microbial Agents

The following microorganisms reported as entomopathogen on cardamom pest. Scientists from Indian Cardamom Research Institute (Spices Board) and Cardamom research station, Kerala Agricultural University has given priority in entomopathogenic microorganisms for pest management in cardamom. The following microorganisms recorded as entomopathogen in cardamom ecosystem.

Name of Pest	Entomopathogens
Cardamom SPCB	*Bacillus thurungiensis* (Stem injection)
Whitefly	*Aschersonia placenta, verticillium* spp
Root grub	*Beauveria basiana, Metarhizium anisopliae, Bacillus poppilae, EPN – Heterorhabiditis* spp, *Sterinernema* spp
Hairy caterpillar	NPV
Nematode	*Paecilomyces lilacinus, Pasteuria penetrans, Trichoderma* sp, VAM – *Glomus fasciculatum, Gigaspora margarita*

(Thomas, 2006)

12. Insecticide Management

- *Thrips*

  Attempt tot control the trips was made first during the early forties and tobacco decoction spray was found out to be effective for trips control by Cherian and Kailasam (1941). Later, Jones (1943) reported that spraying with 120 ml of nicotine sulphate and 120 ml of fish oil resin soap dissolved in 450 litres of water at weekly spraying effectively controlled the thrips. Subbiah (1949) recommended HCH@ 4 Kg/ ha for effective control of cardamom thrips. Use of nicotine sulphate and HCH were advocated for control of thrips. (Anonymous, 1954; Johne, 1956 and Rajan, 1965).

With the advent of more potent organophorous and carbonate group of insecticides, several field trials were conducted and an array of insecticides recommended to manage thrips damage in cardamom ecosystem Viz; quinalphos 0.025% and 0.05%, monocrotophos 0.03%, fenthion 0.05%, methyl parathion 0.05%, chloripyriphos 0.05%, methidathion 0.05%, carbosulfan 0.05%, However, synthetic pyrethroids application was found to increase thrips damage with regular application.

A schedule of spray for thrips management was fixed as seven rounds for high ranges of Kerala and five rounds for lower Pulneys, Tamilnadu. Profenofos 0.05 percent, Lambda cyhalothrin 15 ppm and diaphenthiuron were recommended for the control of thrips.

Detailed studies on efficacy of neem formulations against cardamom thrips were conducted and the results indicated that neem oil and other commercial neem formulations were not effective in controlling thrips.

- *SPCB*

  Ayyar 1940 recommended prompt collection and destruction of attacked shoot and capsules and removal of attacked plant parts during off season for efficient management. David *et al.* (1964) recommended various insecticides like fenthion, HCH, DDT. Parathion, malathion, methyldemeton, trichlorfan and carbaryl for controlling this on castor. Dimethoate and phosphamidon 0.05 percent are recommended for cardamom in Kerala. Endosulfan, fenthion, phosalone, monocrotophos, quinalphos, carbaryl+ molasses and dimethoate at 0.1 percent were found effective. A strategy of collection and destruction of moths and infested tillers followed spraying monocrotophos 0.075 percent or fenthion 0.075 percent is recommended against SPCB.

- *Root grub*

  Earlier studies have shown that raking up the soil and subsequent application of granular insectidies (phorate @ 30 – 40 g / plant or carbofuran @ 50 g / plant); during May – June and September – October is an effective method. Reported that application of phorate @ 20-40 g/ plant or chlorpyriphos 0.06% effective against grubs. Later, chlorpyriphos at 0.04% was found to give economical control of grubs.

- *Whitefly*

  Spraying neem oil 0.5 % + emulsifier (non – ionic) 0.5 % to 0.75 % on under surface of leaves, (Gopakumar and Kumaresan, 1991) and Acephate 0.075%, ethion 0.1% and triazophous 0.05% were found equally effective against the nymphs.

- *Nematode*

  Soil fumigation of nursery sites can minimize nematode populations. In plantation area, granular formulation s like phorate @ 2.5 -5.0 g a.i/ plant or carbofuran @ 5g a.i/ plant (twice a year) reduced root knot nematodes and increased yield.

13. Conclusion

Cardamom pest management practices should be a harmonious approach in cardamom eco system. Now a days, farmers adopt cultural practices with insecticide management to reduce the crop loss due to pest. The components of pest management practices like trap crop practices, host plant resistance, pheromone traps; augmentation of natural enemies and bio control agents should be strengthened in field scale adoption. Active participation of farmers is to be encouraged in this research cum demonstration practices in their fields. Importance of pest surveillance system is to be encouraged to the farmers' community to have a fixed plot to monitor the pest status in cardamom eco system. It is essential to work out a suitable pest management practices according to various agro climatic condition of cardamom plantation and also based on pest surveillance status.

References

Ali, S.S and M.N.Venugopal, 1992.Interaction between *Meloidogyne incognita* and *Rhizoctonia solani* in damping off and rhizome rot disease of cardamom seedlings. *Nematol. Medit* 20: 65-66.

Ananthakrishnana, T.N. 1984. Bioecology of thrips. Indira Publishing House, Michigan, USA. 233 Pp.

Anonymous, 1954. Pests of cardamom. Memoirs of the Dept. of the Agri. Madras; 998-1001.

Anonymous, 1981. Annual report for 1978. Central Plantation Crops Research Institute, Kasargod, India, 242 pp.

Ayyar T.V.R. and M.S, Kailasam, 1935. A new disease of cardamom (*Elettaria cardamomum*) apparently due to insect damage in south India. *Bull Ent. Res.* 26:359-369.

Ayyar. T.V.R.1935, A new species of Thysanoptera from south India (*Taeniothrips cardamomi*). Bull ent. Res. 26:357-358.

Cherian, M.C and M.S, Kailasam, 1941. Preliminary studies on the cardamom thrips (*Taeniothrips cardamomi* Ramk) and its control. Madras Agric, 29: 365-369.

David, B.V., P.S Narayanaswami and Murugesan, M.1964.Bionomics and control of the castor shoot and capsule borer *Dichocrocis punctiferalis* Guen. In Madras state. *Indian oil seeds.J.* 8: 146-158.

David, B.V and Sundararaj. 1993. Studies on Dialeurodids of India, *Kanakarajiella* gen.nov.

Eapen S.J. 1994. Pathogenicity of root knot nematodes on small cardamom (*Elettaria cardamomum maton*) *Indian J. nematol* .24:31-37.

Gopakumar, B and S.S, Chandrasekar, 2002. Insect pests of cardamom in cardamom: The genus Elettaria(Ed) P.N. CRC 180-200 Pp.

Gopakumar, B and D. Kumaresan.1991. Evaluation of certain insecticides against cardamom whitefly. *Dialeurodes cadamomi* (David and Subr) *Pestolgy*, 15, 2-5.

Gopakumar, B; Kumaresan, D and S. Varadarasn, 1987. Occurrence of flea beetle, *Basilepta (Nodostoma) fulvicorne* (Jacoby) (Eumolpidae: coleopteran) in cardamom and preliminary studies on its management. *J. coffee Res.*17 (suppl), 154-155.

Johnes, T.J, 1956. Do thrips bother your cardamom, *Indian farming*, 6, 37.

Jones, S, 1943. Control of cardamom thrips (*Taeniothrips cardamomi* Ramk). *Planters' chronicle*, 38, 372-374.

Nair, M.R.G.K, 1978. Cardamom In: A monograph on crop pests of Kerala and their control, Kerala Agricultural University, Vellanikara, 65-74.

Rajan, S.V, 1965. Protect your cardamom against insect pests. *Indian Farming*, 14, 21-24.

Selvakumaran,S and D. Kumaresan, 1993.Studies on bio - ecology, damage potential and control of the whitefly, *Dialeurodes cardamomi* David and Subr. (Aleyrodidade: Homoptera) on small cardamom. Final report – ICAR Scheme. Indian Cardamom Research Institute, Spices Board, Myladumpara, India, 40 p.

Subbiah, M.S.1949.Cardamom thrips and its control. *Madras Agric. J.* 36, 213-215.

Varadarasan, S., R. Manimeghalai, T. Sivasubramonian and D. Kumaresan, 1990. Integrated management of cardamom root grubs. *Spice India,* 3.9-11.

❑❑❑

8

Management of Coconut Pests under Organic Farming Conditions

RAMARAJU, K.
Professor, Department of Agricultural Entomology
Tamil Nadu Agricultural University, Coimbatore 641 003

1. Pest Scenario in Coconut

The coconut palm is subjected to attack by a large number of insect and non-insect pests. More than 800 species of pests have been reported on the crop in India (Kurien *et al.*, 1979). Though coconut palm is infested by a number of insect and non-insect pests in India, only a few pests like Rhinoceros beetle (*Oryctes rhinoceros* L.), Red weevil (*Rhynchophorus ferrugineus* Oliv.), Leaf eating caterpillar (*Opisina arenosella* Wlk.) and White grub (*Leucopholis coneophora* Burm.) have been considered to be the major pests. In recent years, the mites from the sub-order Acari are gradually raised in numbers and caused extensive damage and loss to coconut. Nine species of eriophyid mites have been recorded to infest coconut fronds and nuts from different parts of the world. A mong them, *Aceria guerreronis* Keifer, popularly known as nut infesting eriophyid mite or perianth mite has proved to be one of the most important pests in the coconut growing countries of the world. Coconut eriophyid mite (CEM) *A. guerreronis* is considered to be one of the very serious pests and found to cause heavy damage resulting in the loss of production of nuts. Generally, the mite was supposed to be present in lower level and becomes alarming only when its population exploded due to some of mysterious ecological factors (Howard *et al.*, 1990). In the recent years, the pest scenario in coconut ecosystem has changed to a little extent with the emergence of a few newer pests. Further, frequent occurrence of coccids and slug caterpillars, which are hitherto considered as minor pests, was reported in a few locations in Tamil Nadu.

In general, most of the coconut pests are being managed by Integrated Pest Management strategies. An IPM package includes the judicious combination of mechanical, biological and chemical methods. Among the IPM components, biological control has been adopted successfully against the rhinoceros beetle and the leaf eating caterpillar. But chemicals are predominantly used for the control of red palm weevil and indiscriminate use of insecticides has been reported in recent years for the management of eriophyid mite, slug and leaf feeding caterpillars in many areas.

In early days, coconut cultivation was done in coastal areas without application of any fertilizers and pesticides. The area under coconut has been increased tremendously during the past three decades due to several reasons and in many places farmers adopted intensive cultivation methods to increase the production. The era of Green Revolution of 1960s gifted the Indian Agriculture with fertilizer responsive high yielding varieties and highly effective fast acting pesticides boosted the production to greater heights. In coconut, chemical fertilizers were also applied to attain maximum productivity. Considering the nature of the palms (height of the trees), the feasibility of applying pesticides was not done extensively. Hence, more emphasis was given for the adoption of biological control methods to contain many major pests and diseases, since 1960. But in recent years, some insecticides and fungicides were used regularly in many gardens to manage the out break of major pests like red palm weevil; eriophyid mite, black headed caterpillar and basal stem rot and wilt diseases. However, indiscriminate use of pesticides combined with favourable weather factors has triggered the secondary out breaks of a few minor pests *viz.*, slug and leaf feeding caterpillars and mealy bugs etc. in one or two places in Coimbatore, Erode and Kanyakumari districts in Tamil Nadu, which was reported by Swamiappan and SundaraBabu (1995) and Ramaraju *et al.* (2007a). Now after realizing the ill effects of chemical fertilizers and pesticides a few farmers is adopting 'organic farming or Do nothing farming' methods for coconut cultivation. Today in India, as practiced elsewhere, Organic Agriculture is gaining momentum, in which use of bio inputs plays a predominant role to sustain the production and productivity. The already available IPM packages can be very well be used for the effective management of major coconut pests under organic farming conditions also.

2. What is Organic Farming?

Organic farming is a system which avoids or excludes the use of synthetic inputs. Organic food comes from organic farms utilizing the best of both traditional and modern techniques. It is a combination of techniques to build soil fertility for sustainable production, mostly using local, and natural resources with least external inputs. Organic agriculture is a unique production management system which promotes and enhances agro-ecosystem health, including biodiversity, biological cycles and soil biological activity, and this is accomplished by using on-farm agronomic, biological and mechanical methods in exclusion of all synthetic off-farm inputs (FAO).

2.1 Do Nothing or Natural Farming

The 'Do nothing farming' is the other name of 'natural farming'. i.e. The farming devoid of ploughing, weeding, application of chemicals, pruning and use of minimum water for irrigation.

2.2 Pest Management in Organic Farming

Organic farming is a production system which prevents the use of inorganic inputs. Most of the traditional practices in organic farming can contribute to stable pest management. Often avoiding the use of chemical pesticides by itself will play a major role in containing the pest problems. Several non chemical pest management strategies can be adopted as part of IPM in organic agriculture. The organic methods of pest management in coconut includes 1) ecology based pest management 2) habitat diversification 3) mechanical 4) biological methods / behaviour modifying chemicals and 5) use of botanicals.

2.2.1 Ecology Based Pest Management

The knowledge on the interaction of pest, natural enemies and environment is essential for effective pest management. When the biotic balance in nature is disturbed pest out breaks occur. For example, sudden outbreaks of eriophyid mite and slug caterpillar were reported in Tamil Nadu.

2.2.2 Habitat Diversification and Sanitation

It makes the agriculture environment unfavorable for insect pest population multiplication and establishment. For eg. Grow vanilla, nutmeg, cocoa and banana as inter crops wherever possible. Avoid cultivating arecanut in coconut gardens, since it favours the multiplication of pests as alternate host. Habitat diversification is normally achieved through various practices like crop density, tillage, water management, fertilizer management, organic mulches, sanitation etc. High level of N fertilizer makes palms more susceptible to eriophyid mite. According to Ramaraju *et al.* (2007b) application of high dose of K, farm yard manure and neem cake significantly reduce the mite incidence. Restricting and altering the condition of the breeding sites of beetles, which is the source of larval food supply, proliferation of the pest could be checked.

2.2.3 Mechanical Method / Behaviour Modifying Chemicals

Rhinoceros beetles can be extracted using a beetle hook during peak periods of pest abundance (June to September) from the crown of the palm. The difficulty in advocating chemical methods for the control on rhinoceros beetle and red palm weevil stress the need for the use of behaviour modifying chemicals like pheromones to combat the pests. Since this method is more effective in monitoring pest population and mass trapping. Placing a single synthetic pheromone traps per hectare (like SIME RB and Ferrolure+) was found to be effective in attracting more number of adults.

2.2.4 Biological Methods

The green muscardine fungus *Metarhizium anisopliae* M. is a pathogen, which kills black headed caterpillar in conditions of low temperature and high humidity. The viral pathogen Baculovirus of *Oryctes* is very effective and kills the grub in 15 to 20 days of infection and it affects the longevity and fecundity of the adult beetles. Release of virus inoculated beetles @ 10 – 15 beetles/ha. effectively reduces the population (Mohan *et a.l,* 1989; Biju *et al.,* 1995). Release of larval parasitoids like *Bracon hebetor* Say (Braconidae) *Goniozus nephantidis* Mues. (Bethylidae), prepupal parasitoid, *Elasmus nephantidis* Roh. (Elasmidae) and the pupal parasitoids, *Brachymeria nosotoi* H. (Chalcididae), *Trichospilus pupivora* Ferr. (Eulophidae) were found effective. In addition, conservation of naturally occurring predators also helps to check the pest population under control.

2.2.5 Use of Botanicals

Eco friendly botanicals like castor and neem cakes and botanical pesticides like azadirachtin 1 %, neem oil + garlic soap mixture (2%), Fish oil rosin soap (2.5%) have been proved effective against beetles, leaf eating caterpillars and eriophyid mite in coconut garden. The major pests and their management strategies are briefly summarized below.

3. Rhinoceros Beetle, *Oryctes rhinoceros* (Dynastidae, Coleoptera)

Nature of Damage

Adult beetle bores into the unopened spindle leaves and spathes. The affected leaves when fully open will show characteristic geometric cuts.

Biology

Grubs are stout, sluggish and white in colour with a pale brown head and is usually found in the manure pits at a depth of 5 - 30 cm. Adult beetle is stout, black in colour, about 15 cm long and has a long horn projecting dorsally from the head in male and the horn is smaller in female. The beetle breeds in decaying organic matter, cattle dung heaps, compost pits, felled coconut logs and stumps, coir dust, saw dust heaps, etc. Peak emergence of the adult beetle occurs during the months of June to August. Total life cycle is completed in 3 – 9 months.

Management

- Removal and destruction of all dead coconut trees in the garden to maintain good sanitation.
- Collection and destruction of various bio-stages of the beetle from the manure pits.
- Incorporation of the entomopathogenic fungus, *Metarhizium anisopliae* in manure pits to check the perpetuation of the pest.
- Soaking of castor cake @ 1 kg in 5 litres of water in small mud pots and keeps them in the coconut gardens to attract and kill the adults.
- Longitudinally split tender coconut stem and green petiole of fronds can be treated with fresh toddy and keep them in the garden to attract and trap the beetles.

- Hooking out and killing the adults.
- Setting up of light traps following the first rains in summer and monsoon period to attract and kill the adult beetles.
- Field release of Baculovirus inoculated adult rhinoceros beetle reduces the leaf and crown damage caused by this beetle.
- Mixture of either neem seed powder + sand (1:2) @ 150 g per palm or Neem seed kernel powder + sand (1:2) @ 150 g per palm applied in the base of the 3 innermost leaves in the crown effectively controlled rhinoceros beetle damage.
- Setting up of Rhinolure @ 1 / ha. to attract adult beetles.

4. Red Palm Weevil, *Rhynchophorus ferrugineus* (Curculionidae, Coleoptera)

Nature of Damage

Presence of holes, oozing out of viscous brown fluid and extrusion of chewed up fibres through the holes, longitudinal splitting of leaf bases and withering of central spindle. In severe and later stages of infestation the sound of gnawing and breaking of fibres could be heard in silent environment.

Biology

The reddish brown weevil has six dark spots on thorax and in male the conspicuous long snout has tuft of hairs. The female lays the eggs in scooped out small cavities on palms. Grubs are apodous and light yellowish in colour with a red head. Pupation takes place in a fibrous cocoon inside the trunk.

Management

- Removal and burning of damaged plants in coconut gardens to prevent further perpetuation of the pest.
- Injuries on stems of palms should be avoided as the wounds may serve as oviposition sites for the weevil and all the holes should be filled with cement.
- Cutting of green leaves should be avoided and if needed, they should be cut about 120 cm. away from the stem.
- Setting up of attractant traps (mud pots) containing sugarcane molasses 2.5 kg/toddy 2.5 litres + acetic acid 5 ml + yeast 5 g +

longitudinally split tender coconut stem/logs of green petiole of leaves of 30 numbers in one acre to trap adult red palm weevils in large numbers.

- Root feeding of azadirachtin 5%10 ml mixed with 10 ml of water.
- Setting up of Ferrolure @1 / ha. to trap adult beetles.

5. Black headed caterpillar, *Opisina arenosella* (Cryptophasidae, Lepidoptera)

Nature of Damage

The caterpillars construct galleries out of silken webs reinforced with the excreta and the scraps of leaf bits. They remain on the lower surface of the coconut leaflets in silken galleries and devour the chlorophyll containing tissues. The affected portion gets dried presenting a scorched up appearance to the foliage. As the caterpillars feed on the chlorophyll containing tissues, the photosynthetic efficiency of the palm is reduced drastically resulting in poor nut production.

Biology

Creamy white coloured, oval shaped eggs are laid by the female moths on the lower side of the leaflets on or near the old larval galleries. Larva is greenish brown in colour which is cylindrical and slightly compressed with a posterior tapering end. Head of the larva is black in colour and brown stripes are present on the body. Pupa is dark brown in colour with blunt anterior and tapering posterior ends. Adult is a greyish white small moth and the females will be possessing a long antenna and three faint spots on forewings. Male moths will be smaller than the female with a slender abdomen and a conspicuous tuft of hairs at the base of the hindwing. The pest completes its life cycle from egg to adult in about two months.

Management

- Removal and destruction of all affected leaves/leaflets.
- Release of larval parasitoids namely, *Goniozus (Parasierola) nephantidis* @ 3,000/hectare under the coconut trees periodically from January at fortnightly intervals will check the build up of the pest during summer.

- Spraying of azadirachtin 1% to cover the under surface of the leaves thoroughly in case of severe epidemic outbreak of the pest in young palms.
- Root feeding of azadirachtin 5% 10 ml mixed with 10 ml of water.

6. Leaf Caterpillar, *Turnaca acuta* (Notodontidae, Lepidoptera)

Nature of Damage

Grown up caterpillars feed on coconut leaves leaving the midrib as sticks. Young larvae were always found feeding at the terminal end of mostly young leaflets.

Biology

Young caterpillars are brown in colour with two pink coloured stripes, whereas, the grown up caterpillars are green in colour with swollen head. A pink coloured 'V' shaped marking is seen on the head and two longitudinal pink coloured stripes are found on the larval body. The larval period will be about 20 days and they pupate in leaf fold made by webbing. Pupal period will last for 10 days and the adult moths are dull white in colour with bipectinate antennae and bulge abdomen. Adults lay eggs in groups firmly glued on the leaves and are arranged in 1-2 rows containing 7-13 eggs. Eggs are very hard and round in shape with brown mottlings and ring like markings.

Management

- Removal of the affected leaves in the lower rows.
- Monitoring the pest occurrence with the help of light traps.
- Spraying of nee oil @ 3 0 ml/litre of water with a rocker sprayer.

7. Slug Caterpillar, *Contheyla rotunda* (Limacodidae, Lepidoptera)

Nature of Damage

Larvae feed by scrapping the green leaf tissues which subsequently results in drying and feeding holes on the dried portion. Affected leaf fronds show burnt-up appearance.

Biology

The caterpillar is greenish brown, slug-like, with broader anterior and narrow posterior region. It has tiny bunches of minute urticating hairs arranged in rows all over the body on the sides and dorsum. The larval period lasts for 25-45 days. The larvae pupate inside a round white shell like cocoon attached on leaves. Female moths are larger with a filiform antenna and males are smaller with a bipectinate antenna. They lay about 200 eggs in groups on leaves.

Management

- Removal of the affected leaves in the lower rows.
- Monitoring the pest occurrence with the help of light traps.
- Spraying of neem oil @ 30 ml/litre of water with a rocker sprayer.

8. Scale, *Aspidiotus destructor* (Diaspididae, Hemiptera)

Nature of Damage

These insects usually cluster in large numbers on the underside of the leaflets, some times also in the fleshy basal stalk portion of the compound inflorescence and the inner side of the perianth. In severe outbreaks, the insect multiplies with abundance so that it appears like a continuous yellow crust all over the underside of the leaflets. It causes premature death of the leaflets by sucking their sap and obstructing the stomatal opening. Infested leaves turn yellow in colour, wither and die. Recently, Ramaraju *et al.* (2007a) reported that the application of fenpropathrin caused the secondary outbreak of scales, *Pseudaulacaspis cockerelli* Cooley infesting leaves and nuts.

Biology

It is a small, flat, whitish scale with a waxy covering. Females are circular in outline and males are oval. Eggs are laid under the scale cover and hatch in about eight days giving rise to active young ones. Young ones are small, oval and pale yellow in colour. They affix themselves on favourable tissues by means of rostra and cover themselves individually with a transparent filmy, waxy material. Female lays about 90 eggs in batches at regular intervals. Total life cycle is 32 days for male and 35 days for female.

Management

- Irrigate coconut garden regularly at short intervals.
- Spraying of neem oil @ 30 ml or fish oil rosin soap 2.5 % covering the crown and leaves of the tree checks this pest effectively.

9. Mealy Bug, *Palmicultor palmarum* (Pseudococcidae, Hemiptera)

Nature of Damage

Mealy bugs form small colonies and suck the sap from the spear leaf, spadex, inflorescence and perianth of tender nuts. As a result, the tender unopened fronds remain highly stunted, deformed and suppressed.

Biology

Mealy bugs are oval in shape and measure about 2-3 mm. They are soft bodied with a waxy, white, powdery covering.

Management

- Spraying of neem oil @ 30 ml or fish oil rosin soap 2.5 % covering the crown and leaves of the tree checks this pest effectively.

10. Eriophyid Mite, *Aceria guerreronis* (Eriophyidae, Acarina)

Nature of Damage

Mites infest and develop on the meristematic tissues of growing nuts under the perianth by desapping the soft tissues. Initial symptoms appear in the form of small, pale white or yellow, inverted triangular patches just below the perianth. As the feeding activity increases, it results in physical damage leading to necrosis. In the severely damaged condition, brown patches appear. As the nuts mature, longitudinal fissures and splits occur on the outer surface of the husk. Occasionally, brownish gummy exudate oozes out from the fissures of the nuts. Severe infestation leads to malformed nuts with hardened husk, resulting in reduced copra and fibre yield. Mites occur in large number in two to six months old buttons. Palms of all ages and nut colour are affected by this mite. All the coconut varieties/germplasms are susceptible to this mite. In the severely infested nuts, the husk becomes thick and hardened, making dehusking difficult, besides drastic reduction in fibre yield. There was drastic reduction in both quality and quantity of coir.

Biology

Life cycle of this mite includes egg, two larval instars and an adult stage. Eggs are shiny white and globular in shape which hatch into larvae (protonymph) in three days. The second instar larva is called nymph which subsequently moults into adults. Adult mite is very minute, measuring 200-250 m in length and 35-50 m in width, with two pairs of legs found in the anterior region of the body. It is pale in colour with elongate body and worm-like in appearance. Both nymphs and adults cause damage. Each female can lay approximately 30-50 eggs. It takes about 7-9 days to complete one life cycle from egg to adult. Each colony contains hundreds of eggs, larvae, nymphs and adults. Mite population is usually higher during summer months. The dispersal of mite is primarily through wind.

Management

Several studies have been conducted on the management of coconut mite throughout the country. In general healthy/well maintained gardens show low incidence of mite when compared to poorly/neglected gardens. Application of FYM, neemcake and vermicompost along with botanical spraying significantly reduced the mite damage (Ramaraju *et al.*, 2007b). Do nothing or natural farming of coconut is being practised by a few farmers in Pollachi block, where the mite incidence was significantly low when compared to normal farming methods. Based on the experimental results TNAU has developed a new Integrated Management Package comprising plant nutrients and botanicals has been developed for the management of this mite which includes,

- Application of recommended doses of Urea and Super phosphate and an increased dose of Muriate of Potash to increase the plant resistance to the mite infestation.
- Urea-1.3 kg; Super Phosphate-2.0 kg; Muriate of Potash-3.5 kg/ tree/year.
- Application of well decomposed Farm Yard Manure @ 50 kg and neem cake @ 5 kg/tree/year.
- Soil application of micronutrients: Borax -50 g/tree/year; Gypsum-1.0 kg/tree/year; Magnesium sulphate - 0.5 kg/tree/ year.
- Basin cultivation of green manures like sunnhemp, cowpea, calapagonium etc. and incorporating *in situ*.
- Judicious irrigation and mulching with coconut leaves and husk in the basin.

- Spot (Topical) application of the following eco-friendly agents on the nuts, preferably, during non-rainy season at least three times a year.

Round	Ecofriendly pesticides	Quantity/tree	Period
1	Azadirachtin 1%	5 ml in one lit. of water	January-February
2	Neem oil + Teepol	30 ml in one lit. of water	March - April
3	Azadirachtin 1%	5 ml in one lit. of water	May - June

Note: Besides the above three rounds of spraying two more rounds are required to get maximum control of mites.

- Root feeding can be adopted wherever spraying is difficult. Azadirachtin 1%
- 10 ml+10 ml of water/ tree. Spraying is more effective than root feeding
- Method of application
- Botanicals should be applied in sequence at 45 days interval using a one litre hand sprayer. Rocker sprayer and pedal sprayer can be used for spraying small trees from the ground.
- Spray should be concentrated at the crown region covering only the top six bunches during non-rainy season.
- Bunches must be covered very well by the spray fluid and approximately one litre of spray fluid may be required per tree for spot application.

11. Conclusion

By proper integration of the above nonchemical methods the pests can be managed effectively in a sustainable manner without causing any ill effects to environment.

References

Biju, B., K.Sudhadevi, T.K. Dangar and B.Sathiamma. 1995. Biological suppression of *Oryctes rhinoceros* by re-releases of *Baculovirus oryctes* in an infested contiguous area. *J. Plantn Crops* 23: 62-63.

Howard, F. W., E. Abreu-Rodriguez and H.A. Denmark. 1990. Geographical and seasonal distribution of the coconut mite, *Aceria guerreronis* (Acari: Eriophyidae), in Puerto Rico and Florida, USA. *J. Agric. Univ. Puterto Rico.*, 74: 237-251.

Kurien, C., B. Sathiamma, G.B.Pillai and K.N. Ponnamma. 1979. Insects and mites associated with the coconut palm, (*Cocos nucifera* L.). In: Nematodes, fungi, insects and mites associated with the coconut palm. CPCRI Tech. Bull. 2:93-236. Central Plantation Crops Research Institute, Kasaragod.

Mohan, K.S., S.P. Jayapal and G.B. Pillai. 1989. Biological suppression of coconut rhinoceros beetle (*Oryctes rhinoceros* L.) in Minicoy, Lakshadweep by *Oryctes* baculovirus - impact on pest population and damage. *J. Plantn Crops* 16(Suppl.) : 163 - 170.

Ramaraju , K., E.V.Bhaskaran and S. Kuttalam. 2007 a. Management of nut infesting eriophyid mite with chemicals and secondary outbreak of coccids on coconut. In: *Proceedings of the National conference on Applied Zoology and Sustainable Development*. 13th - 14th July 2007, Hyderabad. P 42-43.

Ramaraju, K., E. V. Bhaskaran, J. Kavitha, P. Pretheep Kumar. 2007 b. Integrated management of coconut eriophyid mite (Aceria guerreronis Keifer) (Acari: Eriophyidae) *Cord* 23 (2), 58-66.

Swamiappan, M. and P.C. Sundara Babu, 1995. Out break of new leaf caterpillar *Turnaca acuta* Wlk. (Notodontidae ; Lepidoptera) on coconut in Tamil Nadu. *Indian Cocon. J.* 25(11):2-4.

❑❑❑

9

Promoting IPM with Focus on Organic Pest Control

HARIPRASAD, Y.
Professor, Department of Entomology, Faculty of Agriculture
Annamalai Univeristy, Annamalainagar -608002, Tamilnadu

1. Introduction

In recent past a variety of techniques have been developed and refined for controlling different insect pests. What is required is to integrate these techniques and fit them into the frame work of IPM .Though significant progress has been achieved in self sufficiency still there are some lacuna and deficiencies. Real agricultural incomes have been stagnating or declining. Of late, agriculture has been chemicalized excessively. This has led to poor sustainability of farm production, though the value of biopesticides and integrated pest management (IPM) in sustainable agriculture has been well recognized, very little is being adopted at filed level. Agricultural produces, particularly food and feed are contaminated with pesticide residues much above the permissible tolerance limits leading to health risks among people and domestic animals. Several pesticide related problems have become aggravated.

"Integrated Pest Management" (IPM) means the careful consideration of all available pest control techniques and subsequent integration of appropriate measures that discourage the development of pest populations and keep pesticides and other interventions to levels that are economically justified and reduce or minimise risks to human health and the environment. IPM emphasises the growth of a healthy crop with the least possible disruption to agro ecosystems and encourages natural pest control mechanisms."(FAO)

IPM strategies consist of three basic components:

- Prevention of pest build-up through use of appropriate crop cultivation methods.
- Observation of the crop to monitor pest levels, as well as the levels of natural control mechanisms, such as beneficial insects, in order to make the correct decision on the need for control measures.Intervention where control measures are needed
- Integrated Pest Management (IPM) is economical, effective as well as health and environmentally-sensitive.

IPM is a systematic and scientific approach to pest management. Steps of IPM are: proper pest identification

- understand pest life cycles and conditions conducive
- monitor for pest presence, locations and abundance
- establish an action threshold (economic, aesthetic, tolerance) consideration and selection of multiple tactics for pest suppression
- evaluation of results

2. Organic Pest Control Involves the Cumulative Effect of many Techniques, Including

- Allowing an acceptable level of pest damage
- Encouraging predatory beneficial insects to control pests
- Encouraging beneficial microorganisms
- Careful crop selection, choosing disease-resistant varieties
- Planting companion crops that discourage or divert pests
- Using row covers to protect crops during pest migration periods
- Rotating crops to different locations from year to year to interrupt pest reproduction cycles
- Using insect traps to monitor and control insect populations.

Each of these techniques also provides other benefits—soil protection and improvement, fertilization, pollination, water conservation, season extension, etc.—and these benefits are both complementary and cumulative in overall effect on farm health. Effective organic pest control requires a thorough understanding of pest life cycles and interactions.

Organic pest control is similar to integrated pest management in some respects.

IPM is a system of farming designed to be sustainable

- IPM involves using the best combination of cultural, biological and chemical measures, including plant biotechnology
- IPM provides a cost effective, environmentally sound and socially acceptable method of managing diseases, insects, weeds and other pest in agriculture

Pesticides & Alternatives IPM Organic farming methods combine scientific knowledge and modern technology with traditional farming practices based on thousands of years of agriculture. The distinguishing principle is an avoidance of synthetic inputs, such as manufactured fertilizers and pesticides, and for this reason, organic methods are easiest to describe by contrasting them with conventional, agrichemical-based methods.

In general, organic pest control rely on naturally occurring biological processes, which often take place over extended periods of time, and a holistic approach, while chemical-based farming focuses on immediate, isolated effects and reductions strategies. In conventional systems, technologyhybrid seed, synthetic chemicals, high-volume irrigation, mechanizationis used to regulate local conditions.

3. Factors Involve in Successful Organic Pest Control

Crop diversity is a distinctive characteristic of organic farming. Conventional farming focuses on mass production of one crop in one location, a practice called monoculture. This makes apparent economic sense: the larger the growing area, the lower the per unit cost of fertilizer, pesticides and specialized machinery for a single plant species. The science of agroecology has revealed the benefits of polyculture (multiple crops in the same space), which is often employed in organic farming.[2] Planting a variety of vegetable crops supports a wider range of beneficial insects, soil microorganisms, and other factors that add up to overall farm health, but managing the balance requires expertise and close attention.

3.1 Soil Fertility

The central farming activity of fertilization illustrates the differences. Organic farming relies heavily on the natural breakdown of organic matter, using techniques like green manure and composting, to replace nutrients taken from the soil by previous crops. This biological process, driven by microorganisms, allows the natural production of nutrients in the soil throughout the growing season. Organic farming uses a variety of methods to improve soil fertility, including crop rotation, cover cropping, application of compost, and mulching. Organic farmers also use processed natural fertilizers such as bone meal, blood meal, and various mineral powders such as rock phosphate and greensand, a naturally occurring form of potash.

3.2 Pest Control

Differing approaches to pest control are equally notable. In chemical farming, a specific insecticide may be applied to quickly kill off a particular insect pest (animal). Chemical controls can dramatically reduce pest populations for the short term, yet by unavoidably killing (or starving) natural predator insects and animals, cause an ultimate increase in the pest population. Repeated use of insecticides and herbicides and other pesticides also encourages rapid natural selection of resistant insects, plants and other organisms, necessitating increased use, or requiring new, more powerful controls.

4. An IPM System is Designed Around Six Basic Components

- *Acceptable pest levels*: The emphasis is on control, not eradication. IPM holds that wiping out an entire pest population is often impossible, and the attempt can be more costly, environmentally unsafe, and all-round counterproductive than it is worth. Better to decide on what constitutes acceptable pest levels, and apply controls if those levels are exceeded.
- *Preventive cultural practices*: Selecting varieties best for local growing conditions, and maintaining healthy crops, is the first line of defense.

- *Monitoring*: Regular observation is the cornerstone of IPM. Visual inspection, insect traps, and other measurement methods are used to monitor pest levels. Record-keeping is essential, as is a thorough knowledge of the behavior and reproductive cycles of target pests.
- *Mechanical controls*: Should a pest reach an unacceptable level, mechanical methods are the first options to consider. They include simple hand-picking, erecting insect barriers, using traps, vacuuming, and tillage to disrupt breeding.
- *Biological controls*: Natural biological processes and materials can provide control, with minimal environmental impact, and often at low cost. The main focus here is on promoting beneficial insects that eat target pests.
- *Chemical controls*: Synthetic pesticides are generally only used as required and often only at specific times in a pests life cycle. Biological insecticides, derived from plants or naturally occurring microorganisms (eg: Bt), also fit in this category.

IPM is applicable to all types of agriculture. Reliance on knowledge, experience, observation, and integration of multiple techniques makes IPM a perfect fit for organic farming (the synthetic chemical option is simply not considered). For large-scale, chemical-based farms, IPM can reduce human and environmental exposure to hazardous chemicals, and potentially lower overall costs.

Organic cultivation of mixed vegetables in Capay, California. Note the hedgerow in the background

"The role of organic agriculture, whether in farming, processing, distribution, or consumption, is to sustain and enhance the health of ecosystems and organisms from the smallest in the soil to human beings".

International Federation of Organic Agriculture Movement (IFOAM) - The Principles of Organic Agriculture - Principle of health Provides background information and evaluation of the safety and effectiveness of several alternative products and practices that are available for use in insect pest management.It focuses on microbial insecticides; botanical insecticides and insecticidal soaps; attractants and traps; and beneficial insects and mites (natural enemies of pests).

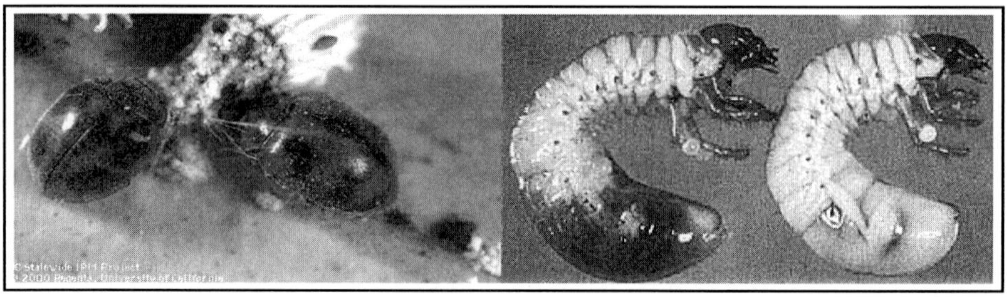

Healthy beetle larva (left); larva infected with *B. popilliae* (right).

Rodolia cardinalis adults attacking cottony cushion scale.

5. Components of Biointensive IPM

Biointensive IPM shares many of the same components as conventional IPM, including monitoring, use of economic thresholds, record keeping, and planning.

How to Get Started with IPM—good planning must precede implementation of any IPM program, but is particularly important in a biointensive program. Planning should be done before planting because many pest strategies require steps or inputs, such as beneficial organism habitat management, that must be considered well in advance. Attempting to jump-start an IPM program in the beginning or middle of a cropping season generally does not work.

5.1 When Planning a Biointensive IPM Program, Some Considerations Include

- Options for design changes in the agricultural system (beneficial organism habitat, crop rotations).
- Choice of pest-resistant cultivars.
- Technical information needs.
- Monitoring options, record keeping, equipment, etc.

IPM options may be considered proactive or reactive. Proactive options, such as crop rotations and creation of habitat for beneficial organisms, permanently lower the carrying capacity of the farm for the pest. The carrying capacity is determined by factors like food, shelter, natural enemies complex, and weather, which affect the reproduction and survival of a species. Cultural controls are generally considered to be proactive strategies.

The second set of options is more reactive. This simply means that the grower responds to a situation, such as an economically damaging population of pests, with some type of short-term suppressive action. Reactive methods generally include inundative releases of biological controls, mechanical and physical controls, and chemical controls.

A crucial step in any IPM program is to identify the pest.

Monitoring (field scouting) and economic injury and action levels are used to help answer these and additional questions.

5.2 Monitoring

Monitoring involves systematically checking crop fields for pests and beneficials, at regular intervals and at critical times, to gather information about the crop, pests, and natural enemies. Sweep nets, sticky traps, and pheromone traps can be used to collect insects for both identification and population density information. Leaf counts are one method for recording plant growth stages. Square-foot or larger grids laid out in a field can provide a basis for comparative weed counts. Records of rainfall and temperature are sometimes used to predict the likelihood of disease infections. Specific scouting methods have been developed for many crops.

5.3 Economic Injury and Action Levels

The economic injury level (EIL) is the pest population that inflicts crop damage greater than the cost of control measures. Because growers will generally want to act before a population reaches EIL, IPM programs use the concept of an economic threshold level (ETL or ET), also known as an action threshold. The ETL is closely related to the EIL, and is the point at which suppression tactics should be applied in order to prevent pest populations from increasing to injurious levels.

5.4 Time and Resources

A successful biointensive IPM program takes time, money, patience, short- and long-term planning, flexibility, and commitment. The pest manager must spend time on self-education and on making contacts with Extension and research personnel. Be aware that some IPM strategies, such as increasing beneficial insect habitat, may take more than a year to show results.

"Rotation crops, when accompanied by care in the use of pure seed, is the most effective means yet devised for keeping land free of weeds. No other method of weed control, mechanical, chemical, or biological, is so economical or so easily practiced as a well-arranged sequence of tillage and cropping."

5.5 Integrated Pest Management Control Tactics

- Pest-resistant crops
- Cultural Control
- Physical and Mechanical Control
- Chemical Control

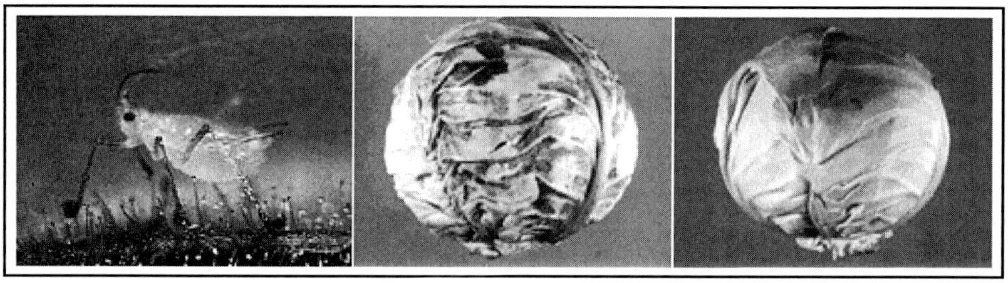

Left : Pest-resistant crops include potatoes bred to express hairy, sticky leaves that can trap and kill aphids Center and Right (respectively): Susceptible cabbage attacked by onion thrips, and a thrips resistant cabbage grown in the same field.

5.6 Pest-Resistant Crops

One of the mainstays of integrated pest management is the use of crop varieties that are resistant or tolerant to insect pests and diseases. A resistant variety may be less preferred by the insect pest, adversely affect its normal development and survival, or the plant may tolerate the damage without an economic loss in yield or quality.. Commercial corn lines with increased concentration of a chemical called DIMBOA display resistance to European corn borer, a major pest throughout the world. In the case of cabbage, the only reliable method of controlling onion thrips is through the use of resistant varieties.

Resistant varieties are continually being bred by researchers. Growers can also do their own plant breeding simply by collecting non-hybrid seed from healthy plants in the field. The plants from these seeds will have a good chance of being better suited to the local environment and of being more resistant to insects and diseases. Since natural systems are dynamic rather than static, breeding for resistance must be an ongoing process, especially in the case of plant disease, as the pathogens themselves continue to evolve and become resistant to control measures.

5.7 Cultural Control

Definition: Manipulation of cultural practices to the disadvantage of pests

Cultural controls are manipulations of the agroecosystem that make the cropping system less friendly to the establishment and proliferation of pest populations. Although they are designed to have positive effects on farm ecology and pest management, negative impacts may also result, due to variations in weather or changes in crop management.

Proactive Strategies (Cultural Controls)

- Healthy, biologically active soils (increasing belowground diversity)
- Habitat for beneficial organisms (increasing aboveground diversity)
- Appropriate plant cultivars

Genetic diversity of a particular crop may be increased by planting more than one cultivar. For example, a recent experiment in China demonstrated that disease-susceptible rice varieties planted in mixtures with resistant varieties had 89% greater yield and a 94% lower incidence of rice blast (a fungus) compared to when they were grown in monoculture.

5.8 Crop Rotation

Crop rotation avoids a decrease in soil fertility, as growing the same crop repeatedly in the same place eventually depletes the soil of various nutrients. A crop that leaches the soil of one kind of nutrient is followed during the next growing season by a dissimilar crop that returns that nutrient to the soil or draws a different ratio of nutrients, for example, rices followed by cottons. By crop rotation farmers can keep their fields under continuous production, without the need to let them lie fallow, and reducing the need for artificial fertilizers, both of which can be expensive.Legumes, plants of the family Fabaceae, for instance, have nodules on their roots which contain nitrogen-fixing bacteria. It therefore makes good sense agriculturally to alternate them with cereals (family Poaceae) and other plants that require nitrates. A common modern crop rotation is alternating soybeans and maize (corn). In subsistence farming, it also makes good nutritional sense to grow beans and grain at the same time in different fields.This principle is of particular use in organic farming, where pest control may be achieved without synthetic pesticides. A traditional component of crop rotation is the replenishment of nitrogen through the use of green manure in sequence with cereals and other crops. It is one component of polyculture. Crop rotation can also improve soil structure and fertility by alternating deep-rooted and shallow-rooted plants

Nematode pests commonly controlled with crop rotation methods include the golden nematode of potato, many root-knot nematodes, and the soybean cyst. A corn/soybean rotation is one example of rotating compatible economic crops

Multiple cropping is the sequential production of more than one crop on the same land in one year. Depending on the type of cropping sequence used, multiple cropping can be useful as a weed control measure, particularly when the second crop is interplanted into the first.

Interplanting is seeding or planting a crop into a growing stand, for example overseeding a cover crop into a grain stand. There may be microclimate advantages (e.g., timing, wind protection, and less radical temperature and humidity changes) as well as disadvantages (competition for light, water, nutrients) to this strategy. By keeping the soil covered, interplanting may also help protect soil against erosion from wind and rain.

5.9 Intercropping

The practice of growing two or more crops in the same, alternate, or paired rows in the same area. This technique is particularly appropriate in vegetable production.

Strip cropping is the practice of growing two or more crops in different strips across a field wide enough for independent cultivation (e.g., alternating six-row blocks of soybeans and corn or alternating strips of alfalfa and cotton or alfalfa and corn). It is commonly practiced to help reduce soil erosion in hilly areas.

5.10 Sanitation

Involves removing and destroying the overwintering or breeding sites of the pest as well as preventing a new pest from establishing on the farm (e.g., not allowing off-farm soil from farm equipment to spread nematodes or plant pathogens to your land). This strategy has been particularly useful in horticultural and tree-fruit crop situations involving twig and branch pests.

5.11 Spacing of Plants

Heavily influences the development of plant diseases and weed problems. The distance between plants and rows, the shape of beds, and the height of plants influence air flow across the crop, which in turn determines how long the leaves remain damp from rain and morning dew.

Altered planting dates can at times be used to avoid specific insects, weeds, or diseases. For example, squash bug infestations on cucurbits can be decreased by the delayed planting strategy, i.e., waiting to establish the cucurbit crop until overwintering adult squash bugs have died. In some cases, "degree days" needed for certain important insect pests to develop. Insects, being cold-blooded, will not develop below or above certain threshold temperatures. Calculating accumulated degree days, that is, the number of days above the threshold development temperature for an insect pest, makes the prediction of certain events, such as egg hatch, possible. University of California has an excellent Web site that uses weather station data from around the state to help California growers predict pest emergence.

5.12 Mulches

Living or non-living, are useful for suppression of weeds, insect pests, and some plant diseases. Hay and straw, for example, provide habitat for spiders. Research in Tennessee showed a 70% reduction in damage to vegetables by insect pests when hay or straw was used as mulch. The difference was due to spiders, which find mulch more habitable than bare ground. Other researchers have found that living mulches of various clovers reduce insect pest damage to vegetables and orchard crops. Vetch has been used as both a nitrogen source and as a weed suppressive mulch in tomatoes in Maryland. Growers must be aware that mulching may also provide a more friendly environment for slugs and snails, which can be particularly damaging at the seedling stage. Mulching helps to minimize the spread of soil-borne plant pathogens by preventing their transmission through soil splash. Red plastic mulch suppresses root-knot nematode damage in tomatoes by diverting resources away from the roots (and nematodes) and into foliage and fruit.

5.13 Biotech Crops

Gene transfer technology is being used by several companies to develop cultivars resistant to insects, diseases, and herbicides. An example is the incorporation of genetic material from *Bacillus thuringiensis (Bt)*, a naturally occurring bacterium, into cotton, corn, and potatoes, to make the plant tissues toxic to bollworm, earworm, and potato beetle larvae, respectively.

5.14 Adjust Planting Dates

By delaying planting until after the first flush of weeds, new weed seedlings can be killed with a light cultivation or flaming. This practice can help deplete the seed bank in the top layer of soil, resulting in reduced competition later in the growing season. This technique is referred to as a false seedbed approach

There are many agricultural practices that make the environment less favorable to insect pests. Examples include cultivation of alternate hosts (e.g., weeds), crop rotation, selection of planting sites, trap crops, and adjusting the timing of planting or harvest. Crop rotation. The severity and incidence of many plant diseases can also be minimized by crop rotation, and selection of the planting site may affect the severity of insect infestations.

5.15 Trap Crops

Trap crops are planted to attract and hold pest insects where they can be managed more efficiently and prevent or reduce their movement onto valuable crops. Adjusting the timing of planting or harvest is another cultural control technique. The earlier planted processing tomatoes grown are far less likely to be infested by the tomato fruitworm than those planted later in the season. It is also important to use pest-free transplants. Some vegetable crop transplants can be infested with insect pests, and growers using these transplants are put at a considerable disadvantage.

Trenches can capture large numbers of dispersing insects (Colorado potato beetles in spring and fall.

6. Traditional Methods of Pest Control

I. Farm Level Pratices

S.No.	Cropping Techniques	Pest Checked
1.	Ploughing	Red hairy caterpillar
2.	Puddling	Rice mealy bug
3.	Trimming and plastering	Rice grass hopper
4.	Pest free seed material	Potato tuber moth
5.	High seed rate	Sorghum shootfly
6.	Rogue space planting	Rice brown planthopper
7.	Plant density	Rice brown planthopper
8.	Earthing up	Sugarcane whitefly
9.	Detrashing	Sugarcane whitefly
10.	Destruction of weed hosts	Citrus fruit sucking moth
11.	Destruction of alternate host	Cotton whitefly
12.	Flooding	Rice armyworm
13.	Trash mulching	Sugarcane early shoot borer
14.	Pruning / topping	Rice stem borer
15.	Intercropping	Sorghum stem borer
16.	Trap cropping	Diamond back moth
17.	Water management	Brown planthopper
18.	Judicious application of fertilizers	Rice leaf folder
19.	Timely harvesting	Sweet potato weevil

II. Community Level Practices

- Synchronized sowing : Dilution of pest infestation (eg) Rice, Cotton
- Crop rotation : Breaks insect life cycle
- Crop sanitation

  a) Destruction of insect infested parts (eg.) Mealy bug in brinjal

  b) Removal of fallen plant parts (eg.) Cotton squares

  c) Crop residue destruction (eg.) Cotton stem weevil

Advantages	Disadvantages
• No extra skill	• No complete control
• No costly inputs	• Prophylactic nature
• No special equipments	• Timing decides success
• Minimal cost	
• Good component in IPM	
• Ecologically sound	

7. Physical and Mechanical Control

Methods included in this category utilize some physical component of the environment, such as temperature, humidity, or light, to the detriment of the pest. Common examples are tillage, flaming, flooding, soil solarization, and plastic mulches to kill weeds or to prevent weed seed germination.

Heat or steam sterilization of soil is commonly used in greenhouse operations for control of soil-borne pests. Floating row covers over vegetable crops exclude flea beetles, cucumber beetles, and adults of the onion, carrot, cabbage, and seed corn root maggots. Insect screens are used in greenhouses to prevent aphids, thrips, mites, and other pests from entering ventilation ducts

The use of physical barriers such as row covers or trenches prevents insects from reaching the crop. Row covers can help prevent early-season damage to cucurbits by cucumber beetles, and plastic-lined trenches are effective in trapping large numbers of dispersing Colorado potato beetles in the spring and fall. Cold storage is also considered a physical control and, although it does not necessarily kill the insect pests, it at least stops their development and further feeding on the stored crop. Other methods include hand picking of pests, sticky boards or tapes for control of flying insects in greenhouses and various trapping techniques.

7.1 Physical Control

Modification of physical factors in the environment to minimise (or) prevent pest problems. Use of physical forces like temperature, moisture, etc. in managing the insect pests.

7.1.1 Manipulation of Temperature

- Sun drying the seeds to kill the eggs of stored product pests.
- Hot water treatment (50-55°C for 15 min) against rice white tip nematode.
- Flame throwers against locusts.
- Burning torch against hairy caterpillars.
- Cold storage of fruits and vegetables to kill fruitflies (1-2°C for 12-20 days).

7.1.2 Manipulation of Moisture

- Alternate drying and wetting rice fields against BPH.
- Drying seeds (below 10% moisture level) affects insect development.
- Flooding the field for the control of cutworms.

7.1.3 Manipulation of Light

- Treating the grains for storage using IR light to kill all stages of insects (*e.g.*) Infra-red seed treatment unit.
- Providing light in storage go downs as the lighting reduces the fertility of Indian meal moth, *Plodia.*
- Light trapping.

7.1.4 Manipulation of Air

- Increasing the CO_2 concentration in controlled atmosphere of stored grains to cause asphyxiation in stored product pests.

7.1.5 Use of Greasing Material

- Treating the stored grains particularly pulses with vegetable oils to prevent the oviposition and the egg hatching. *e.g.,* bruchid adults.

7.1.6 Use of Visible Radiation

Yellow colour preferred by aphids, cotton whitefly : yellow sticky traps.

7.1.7 Use of Abrasive Dusts

- Red earth treatment to red gram : Injury to the insect wax layer.
- *Activated clay* : Injury to the wax layer resulting in loss of moisture leading to death. It is used against stored product pests.
- *Drie-Die* : This is a porous finely divided silica gel used against storage insects.

7.2 Mechanical Control

Use of mechanical devices or manual forces for destruction or exclusion of pests.

A. Mechanical destruction: Life stages are killed by manual (or) mechanical force.

7.2.1 Manual Force

- Hand picking the caterpillars
- *Beating* : Swatting housefly and mosquito
- *Sieving and winnowing* : Red flour beetle (sieving) rice weevil (winnowing)
- *Shaking the plants* : Passing rope across rice field to dislodge caseworm and shaking neem tree to dislodge June beetles
- *Hooking* : Iron hook is used against adult rhinoceros beetle
- *Crushing* : Bed bugs and lice
- *Combing* : Delousing method for Head louse
- *Brushing* : Woolen fabrics for clothes moth, carper beetle.

7.2.2 Mechanical Force

- *Entoletter* : Centrifugal force - breaks infested kernels - kill insect stages - whole grains unaffected - storage pests.
- *Hopper dozer* : Kill nymphs of locusts by hording into trenches and filled with soil.
- *Tillage implements* : Soil borne insects, red hairy caterpillar.
- *Mechnical traps* : Rat traps of various shapes like box trap, back break trap, wonder trap, Tanjore bow trap.

7.2.3 Mechanical Exclusion: Mechanical Barriers Prevent Access of Pests to Hosts

- *Wrapping the fruits* : Covering with polythene bag against pomegrante fruit borer.
- *Banding* : Banding with grease or polythene sheets - Mango mealybug.
- *Netting* : Mosquitoes, vector control in green house.
- *Trenching* : Trapping marching larvae of red hairy catepiller.
- *Sand barrier* : Protecting stored grains with a layer of sand on the top.
- *Water barrier* : Ant pans for ant control.
- *Tin barrier* : Coconut trees protected with tin band to prevent rat damage.
- *Electric fencing* : Low voltage electric fences against rats.

7.2.4 Advantages

Advantages	Disadvantages
• Home labour utilization	• Limited application
• Low equipment cost	• Rarely highly effective
• Ecologically safe	• Labour intensive
• High technical skill not required in adoption	

7.2.5 Appliances in Controlling the Pests

Insect traps can be used to monitor or directly reduce insect populations. They typically use a food, visual lures, chemical attractants and pheromones that do not injure other animals or humans or result in residues on foods or feeds. Visual lures use light, bright colors and shapes to attract pests. Chemical attractants or pheromones are the substances female insects use to sexually attract males to them. Insect traps can be used in an environmentally sound manner in pest management programs instead of pesticides. No insecticides need to be used and beneficial insects rarely become trapped.

Insect traps can be used to determine when pests emerge, how many there are and where they are coming from. The sooner this information is understood, the easier it is to control pests with other measures.

In many cases traps alone have been shown to control pest populations below levels of economic damage.

Flies and wasps are attracted by proteins. Mosquitoes and many other insects are attracted by bright colors, carbon dioxide, lactic acid, floral or fruity fragrances, warmth, and moisture and pheromones.

- *Light traps* : Most adult insects are attracted towards light in night. This principle is used to attract the insect and trapped in a mechanical device.

a) *Incandescent light trap* : They produce radiation by heating a tungsten filament. The spectrum of lamp include a small amount of ultraviolet, considerable visible especially rich in yellow and red. (*e.g.*) Simple incandescent light trap, portable incandescent electric. Place a pan of kerosenated water below the light source.

b) *Mercury vapour lamp light trap* : They produce primarily ultraviolet, blue and green radiation with little red. (eg.) Robinson trap. This trap is the basic model designed by Robinson in 1952. This is currently used towards a wide range of Noctuids and other nocturnal flying insects. A mercury lamp (125 W) is fixed at the top of a funnel shaped (or) trapezoid galvanized iron cone terminating in a collection jar containing dichlorvos soaked in cotton as insecticide to kill the insect.

Center : A yellow sticky card being used to monitor flea beetle populations.

Right : Pheromone traps can be used to monitor pest populations

c) *Black light trap* : Black light is popular name for ultraviolet radiant energy with the range of wavelengths from 320-380 nm. Some commercial type like Pest-O-Flash, Keet-O-Flash are available in market. Flying insects are usually attracted and when they come in contact with electric grids, they become elctrocuted and killed.

- *Pheromone trap* : Synthetic sex pheromones are placed in traps to attract males. The rubberised septa, containing the pheromone lure are kept in traps designed specially for this purpose and used in insect monitoring / mass trapping programmes. Sticky trap, water pan trap and funnel type models are available for use in pheromone based insect control programmes.

- *Yellow sticky trap* : Cotton whitefly, aphids, thrips prefer yellow colour. Yellow colour is painted on tin boxes and sticky material like castor oil / vaseline is smeared on the surface. These insects are attracted to yellow colour and trapped on the sticky material.
- *Bait trap* : Attractants placed in traps are used to attract the insect and kill them. (eg.) Fishmeal trap: This trap is used against sorghum shootfly. Moistened fish meal is kept in polythene bag or plastic container inside the tin along with cotton soaked with insecticide (DDVP) to kill the attracted flies.
- Pitfall trap helps to trap insects moving about on the soil surface, such as ground beetles, collembola, spiders. These can be made by sinking glass jars (or) metal cans into the soil. It consists of a plastic funnel, opening into a plastic beaker containing kerosene supported inside a plastic jar.
- *Probe trap* : Probe trap is used by keeping them under grain surface to trap stored product insect.
- *Emergence trap* : The adults of many insects which pupate in the soil can be trapped by using suitable covers over the ground. A wooden frame covered with wire mesh covering and shaped like a house roof is placed on soil surface. Emerging insects are collected in a plastic beaker fixed at the top of the frame.
- *Indicator device for pulse beetle detection* : A new cup shaped indicator device has been recently designed to predict timely occurrence of pulse beetle *Callosobruchus spp.* This will help the farmers to know the correct time of emergence of pulse beetle. This will help them in timely sun drying which can bill all the eggs.

7.3 Biological Control

This segment includes several paragraphs with general information about biological control and these subsections:

- Conservation
- Augmentation
- Purchase and Release of Natural Enemies
- Biological Controls

Biological control is the use of living organisms—parasites, predators, or pathogens—to maintain pest populations below economically damaging levels, and may be either natural or applied. A first step in setting up a biointensive IPM program is to assess the populations of beneficials and their interactions within the local ecosystem.

Applied biological control, also known as augmentative biocontrol, involves supplementation of beneficial organism populations, for example through periodic releases of parasites, predators, or pathogens. This can be effective in many situations—well-timed inundative releases of Trichogramma egg wasps for codling moth control, for instance. Biological control agents of plant diseases are most often referred to as antagonists

Most of the beneficial organisms used in applied biological control today are insect parasites and predators. They control a wide range of pests from caterpillars to mites. Some species of biocontrol organisms, such as Eretmocerus californicus, a parasitic wasp, are specific to one host—in this case the sweetpotato whitefly. Others, such as green lacewings, are generalists and will attack many species of aphids and whiteflies.

While augmentative biocontrols can be applied with relative ease on small farms and in gardens, applying some types of biocontrols evenly over large farms has been problematic. New mechanized methods that may improve the economics and practicality of large-scale augmentative biocontrol include ground application with "biosprayers" and aerial delivery using small-scale (radio-controlled) or conventional aircraft.

There are three broad and somewhat overlapping types of biological control: conservation, classical biological control (introduction of natural enemies to a new locale), and augmentation.

Conservation

The conservation of natural enemies is probably the most important and readily available biological control practice available to growers. Natural enemies occur in all production systems, from the backyard garden to the commercial field. They are adapted to the local environment and to the target pest, and their conservation is generally simple and cost-effective. With relatively little effort the activity of these natural enemies can be observed. Lacewings, lady beetles, however fly larvae, and parasitized aphid mummies are almost always present in aphid colonies.

When a natural enemy is successfully established it rarely requires additional input and it continues to kill the pest with no direct help from humans and at no cost.

Augmentation

This third type of biological control involves the supplemental release of natural enemies. Relatively few natural enemies may be released at a critical time of the season (inoculative release) or literally millions may be released (inundative release). Additionally, the cropping system may be modified to favor or augment the natural enemies. This latter practice is frequently referred to as habitat manipulation.

An example of inoculative release occurs in greenhouse production of several crops. Periodic releases of the parasitoid, *Encarsia formosa,* are used to control greenhouse whitefly, and the predaceous mite, Phytoseiulus persimilis, is used for control of the two-spotted spider mite.

Lady beetles, lacewings, or parasitoids such as *Trichogramma* are frequently released in large numbers (inundative release). Recommended release rates for Trichogramma in vegetable or field crops range from 5,000 to 200,000 per acre per week depending on level of pest infestation. Similarly, entomopathogenic nematodes are released at rates of millions and even billions per acre for control of certain soil-dwelling insect pests.

Habitat or environmental manipulation is another form of augmentation. This tactic involves altering the cropping system to augment or enhance the effectiveness of a natural enemy. Many adult parasitoids and predators benefit from sources of nectar and the protection provided by refuges such as hedgerows, cover crops, and weedy borders.

Natural enemies can benefit from a source of nectar. Attractive flowers include, from left to right, wild carrot, dill and goldenrod.

Mixed plantings and the provision of flowering borders can increase the diversity of habitats and provide shelter and alternative food sources. They are easily incorporated into home gardens and even small-scale commercial plantings, but are more difficult to accommodate in large-scale crop production. There may also be some conflict with pest control for the large producer because of the difficulty of targeting the pest species and the use of refuges by the pest insects as well as natural enemies.

Examples of habitat manipulation include growing flowering plants (pollen and nectar sources) near crops to attract and maintain populations of natural enemies. For example, hover fly adults can be attracted to umbelliferous plants in bloom.

Release packs for mass reared natural enemies vary in form and function. From left,*Trichogramma* wasps, *Encarsia* wasps and *Orius* bugs.

Purchase and Release of Natural Enemies

Many commercial insectaries rear and market a variety of natural enemies including predaceous mites, lady beetles, lacewings, praying mantids, and several species of parasitoids. Success with such releases requires appropriate timing (the host must be present or the natural enemy will simply die or leave the area) and release of the correct number of natural enemies per unit area (release rate). In many cases, the most effective release rate has not been identified as it will vary depending on crop type and target host density.

Success also requires a healthy and robust natural enemy. This guide does not make specific recommendations about the purchase or release of the commercially available natural enemies, but it does provide essential information about the biology and behavior of most commercially reared species. This information should be helpful in making decisions regarding their use.

7.3.1 Cotesia

In contrast, insecticides that are more selective because they are most effective against insects with certain feeding habits, at certain life stages, or within certain taxonomic groups, are referred to as "biorational" pesticides. These are also known as "least toxic" pesticides.

Because the biorationals are generally less toxic and more selective, they are generally less harmful to natural enemies and the environment. Biorational insecticides include the microbial-based insecticides such as the *Bacillus thuringiensis* products, chemicals such as pheromones that modify insect behavior, insect growth regulators, and insecticidal soaps. When an insecticide is applied, ideally only the target pest(s) should be affected. The goal is to maximize pest mortality while minimizing harm to natural enemies.

Pesticides are the option of last resort in IPM programs because of their potential negative impacts on the environment, which result from the manufacturing process as well as from their application on the farm. Pesticides should be used only when other measures, such as biological or cultural controls, have failed to keep pest populations from approaching economically damaging levels.

If chemical pesticides must be used, it is to the grower's advantage to choose the least-toxic pesticide that will control the pest but not harm non-target organisms such as birds, fish, and mammals. Pesticides that are short-lived or act on one or a few specific organisms are in this class. Examples include insecticidal soaps, horticultural oils, copper compounds (e.g., bordeaux mix), sulfur, boric acid, and sugar esters. Biorational pesticides. Although use of this term is relatively common, there is no legally accepted definition. Biorational pesticides are generally considered to be derived from naturally occurring compounds or are formulations of microorganisms. Biorationals have a narrow target range and are environmentally benign. Formulations of Bacillus thuringiensis, commonly known as Bt, are perhaps the best-known biorational pesticide. Other examples include silica aerogels, insect growth regulators, and particle film barriers.

Particle film barriers. A relatively new technology, particle film barriers are currently available under the trade name Surround® WP Crop Protectant. The active ingredient is kaolin clay, an edible mineral long used as an anti-caking agent in processed foods, and in such products as toothpaste and Kaopectate. There appears to be no mammalian toxicity or any danger to the environment posed by the use of kaolin in pest control. The kaolin in Surround is processed to a specific particle size range, and combined with a sticker-spreader. Non-processed kaolin clay may be phytotoxic..

Sugar Esters. Sugar esters have performed as well as or better than conventional insecticides against mites and aphids in apple orchards; psylla in pear orchards; whiteflies, thrips, and mites on vegetables; and whiteflies on cotton. However, sugar esters are not effective against insect eggs. Insecticidal properties of sugar esters were first investigated a decade ago when a scientist noticed that tobacco leaf hairs exuded sugar esters for defense against some soft-bodied insect pests. Similar to insecticidal soap in their action, these chemicals act as contact insecticides and degrade into environmentally benign sugars and fatty acids after application. Routine use of any pesticide is a problematic strategy.

Botanical pesticides are prepared in various ways. They can be as simple as pureed plant leaves, extracts of plant parts, or chemicals purified from plants. Pyrethrum, neem formulations, and rotenone are examples of botanicals. Some botanicals are broad-spectrum pesticides. Others, like ryania, are very specific. Botanicals are generally less harmful in the environment than synthetic pesticides because they degrade quickly, but they can be just as deadly to beneficials as synthetic pesticides. However, they are less hazardous to transport and in some cases can be formulated on-farm. The manufacture of botanicals generally results in fewer toxic by-products.

Compost teas are most commonly used for foliar disease control and applied as foliar nutrient sprays. The idea underlying the use of compost teas is that a solution of beneficial microbes and some nutrients is created, then applied to plants to increase the diversity of organisms on leaf surfaces. This diversity competes with pathogenic organisms, making it more difficult for them to become established and infect the plant.An important consideration when using compost teas is that high-quality, well-aged compost be used, to avoid contamination of plant parts by animal pathogens found in manures that may be a component of the compost. There are different techniques for creating compost tea. The compost can be immersed in the water, or the water can be circulated through the compost. An effort should be made to maintain an aerobic environment in the compost/water mixture.

8. Conclusions

Organic pest control with IPM practices facilitates a system of agriculture that is ecologically, economically, and socially viable, in the short as well as long term. Rather than standing for a specific set of farming practices, a sustainable agriculture represents the goal of developing a food production system that + yields plentiful, affordable, high-quality food and other agricultural products + does not deplete or damage natural resources (such as soil, water, wildlife, fossil fuels, or the germplasm base)+ promotes the health of the environment

+ supports a broad base and diversity of farms and the health of rural communities + depends on energy from the sun and on natural biological processes for fertility and pest management + can last indefinitely IPM with organic pest control methods share the goal of developing agricultural systems that are ecologically and economically sound. IPM may be considered a key component of a sustainable agriculture system. A premise common to IPM and sustainable agriculture is that a healthy agroecosystem depends on healthy soils and managed diversity. One of the reasons modern agriculture has evolved into a system of large monocultures is to decrease the range of variables to be managed.

❑❑❑

SECTION - II

UTILIZATION OF BIOLOGICAL CONTROL AGENTS

10

Emerging Trends in Improving the Impact Potential of Biocontrol Agents - Case Study of *Trichogramma*

SITHANANTHAM, S
Director, (R&D), Sun Agro Biosystems, Porur, Chennai – 600 116

1. Introduction

Biological control of insect pests is an important strategy in organic crop protection. Parasitoids, predators and pathogens are the major natural enemies of insect pests and play their roles in natural and applied biological control. Egg parasitoids are an important natural enemy group, as they can suppress the pests before they develop into the injurious (mainly larval) life stage. Among the egg parasitoids, *Trichogramma*tid wasps (Fam.: *Trichogramma*tidae; Order: Hymenoptera) are the most commonly used in biological control.

2. Global Scenario

Trichogramma and *Trichogrammatoidea* are the two commonly used *Trichogrammatid* genera. Annually, over ten million hectares of agricultural and forestry crops are protected by release of mass-produced *Trichogramma*. The target pests are mostly Lepidoptera (moths and butterflies). They are used in a wide array of agricultural crops-cereals, legumes, fruits, plantations, vegetables and tuber crops.

China, Russia and Cuba are among the top countries which mass-produce and utilize the *Trichogramma* for local pest control.

3. Indian Scenario-Historical

The research and development (R & D) on *Trichogramma* is less than a century old. Exploratory mass production and early tests on field release commenced during the 1930s on sugarcane borers in the present Karnataka region. During the next three decades occasional and scattered R & D activities were undertaken. It was during the seventies that the momentum for large scale field release and impact assessment activities was given. Since then there has been a steady increase in the awareness and interest in utilizing mass-produced *Trichogramma.*

4. Mass Production Method

Trichogramma can be mass produced on the eggs of a stored product moth- *Corcyra cephalonica.* This factitious host can be mass-reared in trays containing broken cereal grains (mainly pearl millet, sorghum) as larval diet. 'Starter' eggs (about 6-8 thousand, measuring about 0.5cc are placed in the grain medium (3kg/tray). The eggs hatch in 4-6 days. The larvae web the grains and feed from within. They grow up in about 5-6 weeks and pupal stage lasts for about 7-10 days. The moths live for about 4-7 days; mate and the eggs are laid loosely on the mesh kept at the bottom of the oviposition cage. The eggs are collected, cleaned free off the moth's scales, broken body parts and used as medium for *Trichogramma* mass rearing.

Freshly laid eggs of *Corcyra cephalonica* are usually kept for about 30 minutes under UV light, so to inactivate the embryo. The eggs are then spread on paper cards which are coated with a thin layer of dilute *Acacia* gum. After the gum dries, the eggs in cards are kept in cages/jars with adult Trichogramma (at about one adult per 10-20 eggs) for 2 days for parasitization. The parasitized eggs turn black in about 3-4 days. They are then ready for dispatch for field release, as the adults would emerge from them within the next 2-3 days.

The cards containing *Trichogramma* ('Tricho' cards) are tied to leaves or stem of the target crop. It is preferable to tie them in poles at about crop height and protect them from rain or predation (by ants/spiders) by covering with empty ice-cream cups or other suitable disposable materials. Since the adults usually disperse for about 5-10 metres, it is preferable to release them in units of 10×10m; thus the release points should be ideally 100 per hectare or 40 per acre. Usually the dose rate for 'Tricho' cards will be in cc (usually about 10-12 thousand host eggs per cc) or directly as adults (usually 50-150 thousand per acre) per release. The dose rates and number of release vary with the target pest and target crop (See Table 1).

Table 1: *Trichogramma* Release Rates in Different Crops*

Crop	Target pest	Quantity (cc./acre)	Timing of start of release @ (crop age)
Rice	Stemborer	2-4	From early egg laying (about 2-3 weeks after planting)
	leaffolder	2-4	From early egg laying (about 4-5 weeks age)
cotton	bollworms	6-10	From early flowering (about 12-15 weeks age)
sugarcane	Early shootborer	2-4	From early tillering (about 2-3 weeks age)
	Internode borer	4-6	From early node formation (about 2-3 months age)
Tomato/okra	fruitborer	4-6	From early flowering (about 4-6 weeks after planting)
brinjal	Shoot & fruitborer	4-6	From shoot stage (about 3-4 weeks age)
Cabbage/caulif lower	Diamond back moth	2-4	From early vegetative stage (about 2-3 weeks age)
groundnut	Leafroller	2-4	From early vegetative stage (about 2-3 weeks age)

*Generalized: To be refined according to local experience

5. *Trichogramma* Species and Strains

Though there are many species (both native and exotic) of Trichogrammatids available in India, the more commonly mass produced ones are as follows:

Trichogramma chilonis
T. japonicum
T. pretiosum
T. brasiliensis
Trichogrammatoidea bactrae

Among these, *T. chilonis* is the most widely used for several lepidopteran pests occurring among many target crops. *T. japonicum* is regarded as useful for the Schoenobinae moths (*Tryporyza* in sugarcane and rice) because of slightly longer ovipositor which can help reach the egss below the scales covering them *T. brasiliensis / T. prestiosum* are useful for bollworms/fruitborers on selected target crops. *T. bactrae* is promising for *Plutella* on brassicaceous crops.

The use of improved and/or local strains which are adapted to warmer temperatures is preferable, wherever available. There is scope to also use insecticide-tolerant strains. Presently the Project Directorate of Biological Control (PDBC) at Bangalore can provide some of them.

6. Release Timing in Relation to Pesticide Use

It is possible that farmers may need to spray insecticides for controlling other pest groups and/or for disease control in the same target crop in which Trichogramma release is planned. If 'Tricho' cards have been tied already, the spraying to be delayed for 4-5 days, to permit the adults adequate time for field parasitization. Increase any pesticide had to be sprayed urgently before 'Tricho' release, it is preferable to delay the latter for about 7-10 days, to minimize the residual toxicity to the adult *Trichogramma.*

7. Selectivity/Safety of Pesticides

In general, botanical and microbial biopesticides are found to be safer than synthetic pesticides to 'Tricho' adults. Of course, the relative susceptibility of adults can differ with age and feeding status within a population/strain. There are also inter- and intra- species differences in relative susceptibility of Tricho adults to different pesticides. There are standard protocols for such testing of relative safety (developed by International Organization of Biological Control Working Groups), which categorizes the pesticides into different levels of hazarduousness (or safety) to parasitoids, predators, including *Trichogramma* spp.

❏❏❏

11

Biological Control of Pests in Horticultural Ecosystems

KRISHNAMOORTHY, A.
Division of Entomology and Nematology
Indian Institute of Horticultural Research, Bangalore-560 089, India.

1. Introduction

In recent years, increased interests are shown towards biological control following unsatisfactory results with application of insecticides for control of crop pests. Horticultural crops chiefly the fruits provide a stable environment offering good opportunities for biological control in pest management programme. Many of the key pests of fruit crops are sessile in nature, which are easily available throughout the year for attack by natural enemies. Many examples of successful biological control of pests in fruit crops have created interest in the field of biological control. However, there are certain limitations with the use of biological control agents in short duration annual crops like vegetables and ornamentals, where ecosystem is in continuous flux. Strategies under these cropping systems have been changed for effective biological control of pests. These biocontrol agents have proved as a sole method of pest control or formed as an important component of integrated pest management in horticultural crops. Information on biocontrol work wherever on major pests under each major crop was carried out was gathered and provided for easy understanding on the following headings.

2. Biological Control Tactics

Sl. No.	Pests	Biological control agents and their recommendations
1.	TEMPERATE FRUITS San Jose Scale - *Quadraspidiotus perniciosus* (Comstock)	! Aphelinid parasitoid *Encarsia perniciosi* (Tower) ! *Aphytis* sp. (*proclia* group)/ *Aphytis diaspidis* (Howard) @ 2,000 adults / infested tree once ! Coccinellid predators *Chilocorus infernalis* Mulsant and *Pharoscymnus flexibilis* (Mulsant) ! *C. infernalis* (*C.bijugus*) Mulsant @ 10-20 beetles/tree
2.	Plum Scales - *Eulecanium tiliae* (L) & *Eulecanium tiliae coryli* (L)	! The parasitoids *Coccophagus* sp., *Blastothrix sericea* (Dalman) ! Fungus *Rhinocladiella*
3.	Woolly Apple Aphid, *Eriosoma lanigerum* (Hausmann)	! Parasitoid *Aphelinus mali* ! Kashmir valley ! Nilgiris, Kodaikanal and Shillong ! Recolonisation of *A. mali* @ 1000 adults or mummies/infested tree once ! The predators *Coccinella septempunctata* L., *Hippodamia variegata* (Goeze), *Eupeodes confrater* (Wied.) and *Brinckochrysa scelestes* (Banks)
4.	Codling moth - *Cydia pomonella* (L)	! Leh and Kargil area of J.& K ! Weekly releases of *Trichogramma embryophagum* Htg. and *T. cacoeciae pallidum* Meier @ 2000 adults per tree
5.	Indian Gypsy moth - *Lymantria obfuscata* Wlk.	! The egg parasitoid *Anastatus kashmirensis* Mathur and the larval parasitoid *Exorista rossica* Mesnil
6.	Spider mite- *Tetranychus urticae* Koch	! *Euseius finlandicus* (Oudemans
7.	TROPICAL FRUITS GRAPEVINE Mealybugs	! Encyrtid parasitoid *Anagyrus dactylopii* (How.) ! Coccinellid predator *Scymnus coccivora* Ayyar are ! Adult beetles (10 days old) @ 5000/ha are recommended (Babu, 1986: Mani, 1998) ! *Verticillium lecanii* @ 5g/lit.
8.	*Spodoptera litura* (Fab.) & *Helicoverpa armigera* (Huebn.)	! Nuclear polyhedrosis virus / *Bacillus thuringiensis*
9.	Thrips: *Scirtothrips dorsalis* (Hood)	! Entomopathogens

Contd...

10.	GUAVA Guava aphid - *Aphis gossypii* Glover	! Natural enemies *Cheilomenes sexmaculata, Scymnus* sp., *Pseudaspidimerus trinotatus* (Thunberg) (=*P. circumflexa* (Mots.)), *C. nigrita, C. montrouzieri, Ischiodon scutellaris* (F.) and *Paragus serratus* (F), (Mani & Krishnamoorthy, 1989).
11.	Green Shield Scale - *Chloropulvinaria psidii* (Mask.)	! Three parasitoids viz., *Coccophagus cowperii* (Grlt.) ! *C. bogoriensis* (Kon.), *Aneristis* sp. ! the predator *S. coccivora* are known to attack *C. psidii* ! A release rate 20 adult beetles per tree, (Mani and Krishnamoorthy, 1990).
12.	Mealybugs *Ferrisia virgata* (Ckll.)	! The parasitoids *Aenasius advena* Comp., and *Blepyrus insularis* (Cam.) and predators *Scymnus coccivora, Mallada boninensis* (Okamoto), *Brumoides suturalis* (F.) and *Spalgis epius* Westwood
13.	*Planococcus citri*	! The parasitoids *Coccidoxenoides peregrina* (Timberlake), *Allotropa* sp. and the predators *S. epius* and *C. montrouzieri* regulate *P. citri* in nature. (Mani, 1994)
14.	*Planococcus lilacinus* (Ckll.) & *Maconllicoccus hirsutus*	! The predators *S. epius* and *C. montrouzieri*
15.	Wax Scale - *Drepanococcus* (=*Ceroplastodes) chiton* (Green)	! The parasitoids *Anicetus ceylonensis* How. and *Cephaleta brunniventris* (Mani, 1995)
16.	Spiralling whitefly – *Aleurodicus dispersus* Russell	! Natural enemies, *C. montrouzieri, Axinoscymnus puttarudiahi* Kapur & Munshi and *Mallada astur* (Mani and Krishnamoorthy, 1999) ! Two aphelinid parasitoids namely *Encarsia* sp. nr. *haitiensis* (Dozier) and *Encarsia guadeloupae* Viggiani
17.	CITRUS Scale insects: Cottony cushion scale - *Icerya purchasi* Maskel	! Nilgris Hills in 1927 ! *Rodolia cardinalis* Muls
18.	California red scale – *Aonidiella aurantii* (Mask.)	! Parasitised by *Aphytis* spp. ! Recommended @ 10-15/tree ! Releasing of *A. melinus* 2000/tree (Krishnamoorthy, 1993)
19.	Green scale - *Coccus viridis* (Green)	! *Aneristus ceroplastae* How. and *Encyrtus lecaniorum* Mayr. (Tandon, 1985)

Contd...

20.	Mealybugs (i) Citrus mealybug – *Planococcus citri* (Risso)	Encyrtid *Coccidoxenoides peregrina* (Timberlake) ! An exotic parasitoid *Leptomastix Dactylopii*
21.	Mealybugs – *Nipaecoccus viridis* (Mask.) & *Maconellicoccus hirsutus* (Green)	! The Australian ladybird beetle *C. montrouzieri*
22.	Oriental mealybug - *Planococcus lilacinus* (Ckll.)	! The local parasitoid *Tetracnemoidea indica* (Ayyar)
23.	Citrus butterflies – *Papilio demoleus L.* & *Papilio polytes L.*	! Parasitoids namely *Trichogramma chilonis* Ishii, *Telenomus incommodus* Nixon and *Distatrix papilionis* (viereck), (krishnamoorthy and Singh, 1988) ! *Ooencyrtus papilionis* ! Releases of *T. Chilonis* @ 500 adults per tree ! Application of *Bacillus thuringiensis* @ 1g./lit.supress
24.	Psylla – *Diaphorina citri* Kuwayama	! *Tamarixia* (=*Tetrastichus*) *Radiata* (waterston)
25.	Blackfly - *Aleurocanthus woglumi* Ashby	! Kulu, sikkim, kashi Hills, pusa ! *Encarsia divergence* Silv. and *Encarsia merceti* Silv
26.	Leaf miner – *Phyllocnistis citrella* (Staint.)	! Eight species of parasitoids ! *Cirrostichus (tetrastichus) phyllocnistoides* (Narayanan) and *Ageniaspis* sp. (Rao et *al.*, 2001) ! Release of *Mallada boninensis* @ 20-50/plant
27.	*MANGO* Hoppers – *Amritodus atkinsoni* (Lethierrry) & *Idioscopus* spp.	! The reduviid *Irantha armipes* ! The white halo fungus *Verticillium lecanii* ! Lucknow
28.	Fruit flies – *Bactrocera dorsalis* (Hend.)	! Parasitoids ! *Opius*
29.	Stone weevil – *Sternochetus gravis* (Fabr.) & *S. mangiferae* (Fabr.)	! Predated by ants *Oecophylla smaragdina* (fabr.) *Campontus* Sp., *Monomorium* sp. and the mite *Rhizoglyphus* sp. *Beauveria bessiana*

Contd...

30.	Mealybugs *Rastrococcus iceryoides* (Green)	! The parasitoid *Anagyrus pseudococci* Gir. and the predator *Domomyza* (= *Cacoxenus) perspicax* (Knab.) ! The release of the coccinellid predator *C. montrouzieri* @ 50/tree is found
31.	*Drosicha mangiferae* (Green) & *Perissopneumon ferox* Newst.	! Predated by Coccinellid *Rodolia fumida* Muls.and the chrysopid *Brinckochrysa scelestes* (Banks), and parasitised by *Phygadeuon*
32.	Scale insects : Green shield scale - *Chloropulvinaria polygonata* (Ckll.)	! The lycaenid *Spalgis epius* Westwood and the green lacewing *Mallada astur* Banks
33.	*Aspidiotus destructor* Signoret and *Aulacaspis tubercularis* (Newstead)	! *Thompsonisca desantiellus* Shafee (Tandon and Srivastava, 1980) ! Predated by *Chilocorus nigrita, C. circumdatus* and *Scymnus* sp. An aphelinid *Pteropteryx koebeli* (How.)
34.	Leaf webber - *Orthaga eudrusalis* Walker	! *Brachymeria lasus* (Walker), *Tetrastichus* sp., *Pediobius bruchicida* (Rondani), *Hormius* sp., *Cathartoides* sp.and *Tetrastichus* sp.
35.	*POMEGRANATE* Aphid - *Aphis punicae* (Passerini)	! The coccinellids *Scymnus castaneus* Sic, *S. latemaculatus* Motsch., *C. sexmaculata* and syrphid *P. serratus*
36.	Ash whitefly - *Siphoninus phillyreae* Haliday	! The aphelinid *Encarsia azimi* Hayat
37.	Mealybugs -*P lanococcus lilacinus, P. citri, Nipaecoccus viridis & M. hirsutus*	! *S. epius* and *C. montrouzieri* ! The parasitoid *Leptomastix dactylopii* (Mani and Krishnamoorthy, 2001)
38.	Butterflies *Deudorix epijarbas* Moore	! *Telenomus* sp. ! *Telenomus* sp. *nr. cyrus* Nixon, *Anastatus* sp. *nr. kashmirensis* Mathur and *Trichogramma chilonis* Ishii
39.	*Deudorix isocrates* (Fabricius)	! The egg parasitoids namely *Telenomus* sp., *Ooencyrtus papilionis* Ashmead and *Trichogramma chilotraeae*
40.	CUSTARD APPLE Mealybugs - *M. hirsutus, F. virgata, P. citri, P. lilacinus, P. pacificus* Cox. and *N. viridis*	! *C. montrouzieri* and *S. coccivora* are commonly

Contd...

41.	VEGETABLE CROPS TOMATO Fruit borer - *Helicoverpa armigera* (Huebner)	! Larval parasitoids like *Campoletis chlorideae* Uchida, *Eriborus argenteopilosus* Cam., *Carcelia illota* Curran and *Hexamermis* sp. ! *Trichogramma* viz., *T. chilonis* Ishii, *T. brasiliense* (Ashm.), *T. pretiosum* Riley, *T. exiguum* Pinto & Platner, *T. chilotraeae* Nagaraja & Nagarkatti and *T. perkinsi* Girault (Manjunath *et al.*, 1972) ! *T. chilonis* @ 2.5 lakhs parasitoids/ha at weekly intervals ! With adjuvant during the evening hours at weekly Intervals right from the flower initiation ! *Bacillus thuringiensis* Berliner (Dipel) at 0.25-0.5 kg/ha ! The fungus *Nomuraea rileyi* (Farlow) Samson, (Gopalakrishnan and Narayanan, 1989) ! Four applications of *N. rileyi* @ 3.2 x 10^8 spores/ml along with Triton X-100 (0.01%) at weekly intervals
42.	The American serpentine leaf miner - *Liriomyza trifolii* (Burgess)	! The parasitoids *Hemiptarsenus varicornis* (Girault), *Gonotoma* sp., *Chrysonotomyia* and *Cirrospilus variegates* (Masi)
43.	EGGPLANT Shoot and fruit borer - *Leucinodes orbonalis* Guen.	! Parasitised by *Pristomerus testaceus* Morl, *Diadegma apostata* (G.), *Trathala flavoorbitalis* (Cameron), *Eriborus argentiopilosus, Eriborus sinicus* (Holmgren) *Phanerotoma* sp. ! NPV and the fungus *Bipolaris tetramera* (Mckiney) Shoemaker
44.	Other insect pests	! *Pediobius foveolatus Henosepilachna vigintioctopunctata* (Babu and Ignacimuthu, 2001) ! *Bt* (Thuricide) dust (3x10^6 spores/g), (Sundara Babu, 1970) ! Predator *Cryptolaemus montrouzieri,* (Kalita *et al.*, 1998) ! *Eurytoma* sp. ! Predators ! *Chrysopa orestes* Banks, ! *Telenomus* sp. ! *Herpetogramma bipunctalis* (Sundara Babu *et al.*, 1970)

Contd...

"#	CHILLI/CAPSICUM Thrips - *Scirtothrips dorsalis*	! Anthocorid *Orius insidiosus* ! A eulophid parasitoid *Ceranisus* sp and an entomopapthogen *Verticillium lecanii,* ! (Ganga Visalakshi et al., 2004) *Metarhizium anisopliae* and *Beauveria* spp.
"$%	Aphids - *Myzus persicae* (Sulzer) and *Aphis gossypii* Glover	! *Aphidius* sp. and *Aphelinus* sp. ! *Chrysoperla carnea*
"&%	Gall midge - *Asphondylia gennadi*	! *Eurytoma* sp. (Ayyanna and Raghavaiah, 1990)
"!%	Fruit borers - *Spodoptera litura* & *Helicoverpa armigera*	! Two rounds of NPV @ 250 pob/ha (dhandapani and jaya raj, 1989)
"(%	OKRA Fruit borers - *Earias vittella* (Fab.), *E. insulana* Boisd. & *Helicoverpa armigera*	! *Aleiodes aligharensi* (Qadri), *Aleiodes testaceus* Spinola, *Bracon hebetor* say, *bracon greeni* (Ashm.), *Agathis fabiae* Nixon and *Trichogramma* spp. *Aleiodes aligharensi* (Raja *et al.*, 1998) ! Three applications of *Bt* (Dipel) @ 0.5 kg/ha at weekly (Krishnaiah *et al.*, 1981) ! *Bt* (Biolep) @ 2 kg/ha up to 15 days of application (satapathy and Panda, 1997)
#)%	2. Leafhopper - *Amrasca biguttula biguttula* Ishida	! Two mymarids *Arescon enocki* (Subba Rao and Kaur) and *Stethynium empoascae* (Subba Rao), (Kapadia and Mittal, 1995) ! *Anagrus empoascae* Dozier and *Oligosita* sp.
#*%	Petiole leaf miner - *Melanagromyza hibisci* Spencer	! *Eurytoma* sp. (Krishnakumar and Srinivasan, 1988) ! *Eurytomocharis keralensis* Mukerjee
#+%	CABBAGE/CAULIFLOWER Diamondback moth (DBM) - *Plutella xylostella* (L.)	! *Brachymeria excarinata* (Cherian and Basheer, 1938) ! *Cotesia plutellae* (Kurdy) ! *Diadegmafenestrale* (Holmgr.), *Diadromus collaris* (Grav.) and *C. plutellae* (Chandramohan, 1994) ! *Trichogrammatoidea bactrae* (AVRDC, 1996) ! Entomopathogenic fungi, *Paecilomyces farinosus* (Holmskiold) Brown & Smith and Zoophthora *radicans* (Brefeld) were isolated from *P. xylostella.* Application of five rounds of the fungus *P. farinosus* @ $1.7x10^8$ spores/ml + triton X-100 (0.01%) at weekly interval significantly (Gopalakrishnan, 1998; Kennedy *et al.*, 2000)

Contd...

#¦%	Cabbage butterfly – *Pieris brassicae* L.	! Larval parasitoids *Cotesia glomerata* (L.) and *Hypersota ebeninus* Grav. Application of *Bt* (Thuricide) at 3x10^6 spores/mg (Atwal and Singh, 1969)
#!'%	Leaf eating caterpillar – *Spodoptera litura*	! *T. chilonis* (Srivastava and Kushwaha, 1995) ! *Telenomus remus* ! *Chelonus heliopae* Gupta @ 75,000 adults/ha when released at 4 times (Gopalakrishnan and Mohan, 1993)
##%	APHIDS - *Brevicoryne brassicae* L.,*Lipaphis erysimi* (Kalt.)&*Myzus persicae* (Sulzer)	! Eleven species of coccinellids Nine species of syrphids The fungus *Erynia neoaphidis* (Remaudiere & Hennebert)
#$%	ONION: Thrips: *Thrips tabaci*	! Two parasitoids viz. *Ceranisus* sp and *Thripobius* (Krishnamoorthy *et al.*, 2006) Predator *Orius tantillus*
#&%	French beans Spider mites - *Tetranychus urticae*	! Indigenous predatory mite *Amblyseius tetranychivorus* Gupta or exotic *Phytoseiulus persimilis* Athias-Henroit @ 20 adults/ plant (Krishnamoorthy and Mani, 1989)
#!%	ORNAMENTAL CROPS Bud borer-*Helicoverpa armigera*	! *Trichogramma chilonis, Trichogramma achaeae, Trichogrammatoidea bactrae* and *Telenomus* sp. (Manjunath, 1972)
#(%	Other lepidopteran pests Chrysanthemum leaf folder - *Hedylepta indicata* (Fab.)	! *Elasmus indicus* Rohu (Cherian, 1929) *Xanthapimpla pedator* (F.) and *E. indicus*
$)%	China aster semilooper - *Ctenoplusia albistriata* Bremer & Gray	! *Euplectrus xanthocephalus*
$*%	Rose hairy caterpillar-*Euproctis lunata*	! Tachinids *Carcelia corvinoides* (Wulp) and *Exorista larvarum* (L.)
$+%	Jasmine bud borer- *Hendecasis duplifascialis* (Hmps.)	! *Cantheconidea furcellata*
$¦%	Jasmine leaf webber – *Nausinoe geometralis* Guenee	! *Chrysoperla carnea* ! *Bacillus thuringiensis*
$!'%	Scales -	! a) *Aonidiella aurantii* b) *Diaspis boisduvalli* Signoret ! *Parasaissetia nigra* (*Niet.)* ! *Coccus hesperidum* L. ! *Cerococcus hibisci* ! The encyrtid parasitoid *Metaphycus luteolus* Timb. *Scutellista caerulea* ! Eulopid *Syntomosphyrum cerococci* Aphelinid *Aneristus ceroplastae* *Pharoscymnus horni* Was found to be an important predator

Contd...

$#%	Aphids -	! *Macrosiphum rosae* Linn.& *M. rosaeiformis* *Macrosiphoniella sanborni* *Aphis* sp. ! *Myzus persicae* Sulz. ! *Aphis gossypii* Glover *Cheilomenes sexmaculata* ! *Toxares macrosiphophagum* Westwood ! *Chrysoperla carnea* ! *Praon aguti* (Sekar, 1960) Coccinellid predators *Coccinella* ! *septempunctata* L. ! *Oenopia sexareata* Muls., *Oenopia kirbyi* Muls. and *Propylea luteopustulata* ! Syrphid predators *Allograpta javana* (Wied.), *Betasyrphus serarius* Wied. and *Sphaerophoria scripta* L.
$$%	Spider mites -	! *Tetranychus urticae* Koch ! *Tetranychus neocalidonicus* Andre ! *Tetranychus ludeni* Zacher ! Eutetranychus orientalis (Klien)
$&%	Mealy bugs -	*Maconellicoccus hirsutus* *Pseudococcus sp.* *Planococcus minor (Mask.)* *Planococcus longispinus Targioni-Tozzetti* *Ferrisia virgata (Ckll.)* Predator *Cryptolaemus montrouzieri*
$!%	Spiralling whitefly - *Aleurodicus dispersus* Russell.	Two aphelinid parasitoids namely *Encarsia* sp. nr. *haitiensis* (Dozier) and *E. guadeloupae* Viggiani
$(%	Thrips- a) *Rhipiphorothrips cruentatus* Hood (Rose)	Predatory thrips *Holopothrips anacardii* (Dhaliwal, 1975) *Polistes hebraeus*
&)%	b) *Microcephalothrips abdominalis*	*Orius maxidentex* Ghauri and *Orius tantillus* (Motsch), (Muraleedharan and Ananthakrishnan, 1978)

3. Future Thrust

- Outstanding success in the control of crop pests has been achieved with exoitic parasitoids and predators. In India too, ecellent control of the pests like cottony cushion scale, apple wooly aphid and citrus mealybug has been obtained with introduced natural enemies. So far, only a few natural enemies have been introduced into India for the control of Horticultural crop pests. Success in biological control is commensurate with the number of efforts made. So importation of natural enemies should be the major approach for the major approach for the suppression of the horticultural crop pests.

- Inundative releases of indigenous natural enemies, where there is a feasibility of large scale multiplication, should be attempted to control important pests.
- There is ample scope of utilizing egg parasitoids especially *Trichogramma* in the supression of vegetable and ornamental crop pests. Large-scale experimentaions are needed to show the utility of *Trichogramma* species in the supression of these crop pests.
- Microbial pathogens have got excellent scope in the conrol of vegetable and ornamental crop pests, since they give immediate control comparable with chemical control. Hence, the usage of insect pathogens should be encouraged in vegetable and ornamental ecosystem for pest control.
- Application of conventional / broad spectrum insecticides interferes with the activity of naturally of occurring biocontrol agents in horticultural ecosystem. To overcome the situation, non-conventional chemicals like botanicals, biopesticides etc., can be applied in the pest management programmes without affecting the key local natural enemies.
- Integration of different bioconrol agents (parasitoid, predator and pathogen) and also with botanicals, conventional insecticide both at reduced doses has to be done to reduse the pest effectively. It is practicable but economics of this integrated approach has to be taken into consideration in management of horticultural crop pests.
- The key natural enemies can be conserved with the use of safer pesticides. All the commonly used pesticides have to be screened for their safety to the key parasitioids and predators and the information should be made available.
- Biological control is the backbone of integrated pest management. Proven or promising natural enemies have to be included as major componant in IPM programmes.
- The use of pheromones to determine the time of release of parasitoids and predaores and kairomones to increase the efficiency of parasitoids should received the attention.

- Though the information on the natural enemies was available for some horticultural crop pests, still exploratory surveys for many more pests have to be conducted in different parts of our country to facilitate the import and export of parasitoids and predatores.
- Publicity on the use of biocontrol agents in he supression of horticultural crop pests through various media helps to create awareness among the farmers.
- Research institutes and other government organisations are unable to supply natural enemies to the farmes in large numbers due to various limitations. At present, there are very few commercial insectaries established in India. Many more large scale multiplication units should be started in all the states to ensure the timely supply of proven natural enemies to th farmers.

References

Atwal, A.S. and Singh, K. 1969. Comparativie efficacy of *Bacillus thuringiensis* Berliner and some contact insecticides against *Pieris brassicae* L. and *Plutella maculipennis* (Curtis) on cauliflower, *Indian J. ent.*, 31: 361-63.

AVRDC, 1996. Annual progress report of Asian Vegetable Research and Development Centre, Shanhua, Taiwan for the Year 1995, 187 p.

Babu, A. and Ignacimuthu, 2001. Regulation of population buildup of *Henosepilachna vigintioctopuncata* F. (Coleoptera: Coccnellidae) by a larval parasitoid Tetrastichus epilachnae (Kurian) (Eulopidae: Hymenoptera). National symposium on: *Pest Management strategies: current trends and future prospects*" held at Chennai, February, 1-2, 2001.

Babu, T.R. 1986. Population density of grape mealybug *Maconellicoccus hirsutus* (Green) and its control by a predator *Cryptolaemus montrouzieri* Muls. Ph.D. Thesis, APAU, Hyderabad.

Chandramohan, N. 1994. Seasonal incidence of diamondback moth, *Plutella xylostella* L. and its parasitoids in Nilgiris, *J. Biol. Control*, 8: 77-80.

Cherian, M.C. 1929. Life history notes on *Lamprosema indicata* (Pyralidae), a caterpillar pest on chrysanthemum. *J. Bomb. Nat. Hist. Soc,.* 33: 857-860.

Cherian, M.C. and Basheer, M. 1938. *Tetrastichus sokolowskii* Kurdj (Family: Eulopidae), a larval parasite of *Plutella maculipennis* in South India. *Indian Acad. Sci.*, 9: 87-89.

Dhaliwal, J.S. 1975. Polistes hebraeus (Fabricius) preying upon *Rhipiphorothrips cruentatus* Hood (Thripidae: Thysanoptera), *Curr. Sci.*, 44: 368.

Dhandapani, N. and Jayaraj, S. 1989. Efficacy of nuclear polyhedrosis virus formulation for the control of *Spodoptera litura* (Fab.) on chillies. *J. Biol. Conrol*, 3: 47-49.

Ganga Visalakshi P.N., Manoj Kumar, A. and A. Krishnamoorthy, 2004. Epizootics of a fungal pathogen. *Verticillium lecanii* Zim. On thrips palmi. *Insect Environment*. 10.

Gopalakrishnan, C., 1998. Studies on the entomopathogenic fungus *Paecilomyces farinosus* (Holmskiold) Brown and Smith for the control of cabbage diamond moth, *Plutella xylostella* L. Ph.D. thesis, Bangalore University, Bangalore, 218p.

Gopalakrishnan, C. and Mohan, K.S. 1993. Epizootics of *Erynia neoaphidis* (Zygomycetes: Entomophthorales) in field population of *Brevicoryne brassicae* (Homoptera: Aphididae) on cabbage. *Entomon*, 18: 21-26.

Gopalakrishnan, C. and Narayanan, K. 1989. Epizootiology of *Nomuraea rileyi* (Fallow) Samson in field populations of *Helicoverpa* (=Heliothis) *armigera* Hubner (Noctuidae: Lepidoptera). *Curr. Sci.*, 57: 867-868.

Kalita, D.N., Roy, T.C.D. and Khound, J.N. 1998. Coccinellid predators on *Aphis gossypii*. *Insect Envronment*, 3(4): 107-108.

Kapadia, M.N. and Mittal, V.P. 1995. Record of two egg parasitoids of the okra jassid *Amrasca biguttula biguttula* (Ishida). *Gujarat Agric. Univ. Res. J.* 20: 180-181.

Karuppuchamy, P. 1994. Studies on the management of pests of pomegranate with special reference to fruit borer *Virachola Isocrates* Fabr. Ph.D. Thesis, TNAU, Coimbatore, 181 p.

Kennedy, J.S. Sairabanu, B. and Rabindra, R.J. 2000. Entomopathogenic fungi fo the management of diamondback moth *Plutella xylostella* on cauliflower. Paper presented in National symposium on microbials in insect pest management. Loyola College, Chennai, 24-25th Feb. 2000.

Krishnakumar, N.K. and Srinivasan, K. 1988. Seasonality of *Eurytoma* sp. – a parasitoid on okra petiole maggot *Melanagromyza hibisci* (Spencer). *Tropical Pest Management*, 34: 331-332.

Krishnamoorthy, A. 1993. Bionomics and development of integrated management strategy for California red scale, *Aonidiella aurantii* (Maskell). Ph.D. Thesis, University of Agric. Sci., Bangalore, India.

Krishnamoorthy, A. and Mani, M. 1989. Effects of releases of *Phytoseiulus persimilis* A-H in the control of two spotted spider mite of French beans. *J. Biol. Cant.* 3: 33-36.

Krishnamoorthy, A. and Singh, S.P. 1988. Observational studies on the occurrence of parasitoids of *Papilio* spp. In citrus. *Indian J. Plant. Prot.* 16: 79-81.

Krishnamoorthy, A. Ganga Visalakshi, P.N. and Mani, M. 2006. A new record of *Thripobicus* sp. a parasitoid of onion thrips *Thrips tabaci* Linderman in India. Paper presented in *Natl. Sym. Biol. Control of sucking pests in India,* IAT, Bangalore May, 26-27m 2996 p. 33-34.

Mani, M. 1998. Bio-efficacy and management of grapevine mealybug. *Indian Inst. Hort. Res. Tech. Bull.,* No.5, 32 p.

Mani, M. 1994. Effectiveness of the exotic parasitoid *Leptomastix dactylopii* in the control of *Planococcus citri* in guava orchards. *J. Ent. Res.,* 18: 351-355.

Mani, M. 1995. Studies on the natural enemies of the wax scale *Drepanococcus chiton* (Green) (Homoptera, Coccidae) on ber and guava in India. *Entomon,* 20: 55-58.

Mani, M. and Krishnamoorthy, A. 1989. Impact of insect predators in the control of *Aphis gossypii* Glover on guava. *J. Biol. Control.* 3: 128-129.

Mani, M. and Krishnamoorthy, A. 1990. Evaluation of the exotic predator *Cryptolaemus montrouzieri* Muls. (Coccinellidae, Coleoptera) in the suppression of green shield scale. *Chlorophulvinaria psidii* (Maskell) (Coccidae, Hemiptera). Entomon, 15: 45-48.

Mani, M. and Krishnamoorthy, A. 1999. Natural enemies and host plants of spiralling whitefly *Aleurodicus disperses* (Homoptera, Aleyrodidae) in Bangalore, Karnataka, Entomon, 24: 75-80.

Mani, M. and Krishnamoorthy, A. 2001. Biological suppression of the mealybugs *Planococcus citri* (Risso) and *Planococcus lilacinus* (Ckll.) on pomegranate in India. *Indian J. Plant Prot.* (In press).

Manjunath, T.M. 1972. Biological studies on *Trichogrammatoidea armigera* Nagaraja, a new dimorphic egg parasite of *Heliothis armigera* (Hubner) in India. *Entomophaga,* 17: 131-147.

Muraleedharan, N. and Ananthakrishnan, T.N. 1978. Bioecology of four species of *Anthocoridae* (Hemiptera: Insecta) *predaceous* on thrips, with key to genera of anthocorids from India. Records of the Zoological survey of India No. 11, pp.1-32.

Raja, J., Rajendran, B. and Pappiah, C.M. 1998. Management of bhendi fruit borer *Earias vittella* F. Proc. *I Natl. symp. Pest Mgmnt. Hort. Crops,* Oct, 15-17, 1997, Bangalore, pp.118-120.

Satapathy, C.R. and Panda, S.K. 1997. Effect of commercial Bt. Formulations against fruit borers of okra. *Insect Environment,* 3(2): 54.

Sekar, P.S. 1960. Host relationships of *Aphidius testaceipes* (Cresson) and Praon aguti (Smith), primary parasites of aphids. *Can. J. Zool.,* 38: 593-603.

Sundara Babu, P.C. 1970. Comparative efficacy of thuricide against the spotted beetle, *Epilachna vigintioctopunctata* F. on brinjal (*Solanum melongena*). *South Indian Hort.,* 18: 85-88.

Sundara Babu, P.C., Lakhmanan, P.L. and Subramanian, T.R., 1970. A note on the efficacy of the microbial insecticides Bactospeine against certain crop pests. *South Indian Hort.,* 18: 96.

Tandon, P.L. 1985. Spatial distribution sampling technique and population dynamics of cirus green scale *coccus viridis* (Green) on mandarin, Ph.D. thesis, UAS, Bangalore.

Tandon, P.L. and Srivastava, R.P. 1980. New records of parasites and predators of important insect pests of mango. *Entomon,* 5: 243-44.

❑❑❑

12

Mass Production Protocols for Potential Biological Control Agents

CHANDISH R. BALLAL & SUNIL JOSHI
Project Directorate of Biological Control (ICAR), Bellary Road, Bangalore

1. Production of Trichogrammatids

The egg parasitoids are produced for mass application. Various steps are given below.

The eggs of *Corcyra* collected are passed through 15, 30 and 40 mesh sieves and run over a slope of paper to eliminate dust particles. The eggs are treated with UV rays (30-w UV tube for 45 minutes at a distance of 2 feet) to prevent hatching. For large quantity of eggs, multi UV chamber hood can be used to treated about 150 cc of eggs in 45 minutes. The eggs are glued to 'Tricho' cards of 15 x 10 cm which are prepunched to obtain 16 pieces (measuring 4 x 3 cm) leaving uncovered space at one end to facilitate stapling. The eggs are exposed to adult female *Trichogramma* in the ratio of 30:1 for 24 hours. In case cards in polythene bags are exposed, the eggs ratio should be 40:1 but in this method the females are allowed to parasitise till they die. After parasitisation, 5-day-old parasitised egg cards are prepared for shipment. For distant places, 3 days old exposed cards are sent. Each card is stapled in such a way that the eggs do not touch each other.

Twenty or more 'Tricho' cards are packed in each polythene bag depending on size of the bag. In polythene bags, strips of wood wool coated with concentrated and dried honey are placed as feeding for the adult parasitoid, which may emerge in transit.

2. Mass Production of Chrysopids

Chrysopids belong to the order Neuroptera. They are otherwise known as green lacewings. The adults are either predaceous or feed on honeydew and pollen grains. The larvae are predaceous, feeding on the eggs and neonate larvae of lepidopterans, nymphs and adults of whiteflies, aphids and other homopterans. They are also known as aphid lions or aphis lions. They are encountered in most agricultural and horticultural ecosystems. There are about 69 species of chrysopids recorded from India. Among them *Chrysoperla carnea* and *Mallada boninensis* are the most common species.

2.1 Production Protocol

Adults (approximately 200 in number) are kept in an oviposition cage measuring 75 x 30x 30 cm. The sides of the cage are lined with smooth nylon wire mesh of '0 size (not preferred for egg laying) and the sliding top cover is fitted with black cloth for deposition of eggs. To prevent damage to the eggs the top is slid over a comb fitted on one side of the cage. The sliding top cover is replaced on alternate days starting from the 5th day onwards. The oviposition cage is kept for 30- 35 days and the dead adults are removed every alternate day.

The adults in the oviposition cage are fed on alternate days on items provided on swabs kept in plastic plates as detailed below:

Drinking water

50% honey

Protinex mixture (Equal quantity in volume of Protinex + fructose + honey + powdered yeast dissolved in small quantity of water). In rural areas even powdered black gram (with seed coat) can be mixed instead of Protinex.

Castor pollen

The eggs deposited on the black cloth forming the top cover of the oviposition cage are kept for 24 hours to facilitate hardening of chorion and dislodged by gently rubbing with a piece of sponge. Eggs collected can be used for further multiplication or for field release.

In the first step of larval rearing, three-day-old chrysopid eggs numbering 120 are mixed with 0.75 cc of UV inactivated *Corcyra* eggs in a plastic container (27x18x6 cm). On hatching, the larvae start feeding. On the 3rd day, the larvae are transferred to 2nd step individual rearing in 2.5cm cubical cells of plastic louvers. Each louver can hold 192 larvae. *Corcyra* eggs are provided in all the cells of the louver by sprinkling. Feeding is provided in 2 doses - first feeding of 1.50 cc for 100 larvae and second feeding of 2 cc for 100 larvae with a gap of 3-4 days. The total quantity of *Corcyra* eggs required for rearing 100 chrysopid larvae is 4.25 cc. Each louver is secured on one side by brown paper sheet and after transfer of larvae it is covered with acrylic sheet and clamped on all the sides. Brown paper is used for clear visibility of *Corcyra* eggs and for facilitating cocoon formation.

Cocoons are collected after 24 hours of formation (when they get hardened) by removing paper from one side. The cocoons are placed in adult oviposition cages for emergence.

One set of louver remains in use for 13 to 15 days. After utilisation, louvers are cleaned, sterilised and reused.

For field release viable eggs are mixed with *Corcyra* eggs, saw dust and paper strips before sending the shipment. Once the larvae emerge, they are released in the field.

2.2 Production Procedure for Mealybug Feeding Coccinellids *Cryptolaemus montrouzieri* Mulsant

C. *montrouzieri* has been introduced from Australia for the control of *Coccus viridis*. But the predator has established on many species of mealybugs and the green shield scale. In the field its effectiveness was not demonstrated until a programme was initiated at Central Horticultural Experiment Station, Chethalli where detailed studies resulted in finding out the methods of its practical use for the suppression of mealybugs on citrus, coffee, grapes, several other fruit crops and ornamentals.

2.2.1 Production Procedures

After 15 days of infestation of pumpkins with *P. citri* they are exposed to a set of 100 beetles for 24 hrs. and kept back in a cage as described for the production of *P. citri*. The beetles during the period of exposure feed on mealybugs as well as deposit their eggs singly or in groups of 4-12. The grubs are visible in such cages within a week of exposure to beetles. The young grubs feed on eggs and small mealybugs but as they grow they become voracious and feed on all stages of mealybugs. For facilitating the pupation of grubs dried guava leaves or pieces of paper are kept at the base of each of the cages. The first beetle from the cages start emerging on 30th day of exposure to *C. montrouzieri* adults. The beetles are collected daily and kept in separate cages for about 10-15 days to facilitate completion of mating and pre-oviposition. The beetles are also fed on honey agar medium. From each cage about 175 beetles are obtained. The emergence of the beetles is completed within 10 days. The cost of production of 100 larvae at 1977 price index was Rs.6/- and at 1992 Rs.60. The cost of production of 100 beetles at 1977 price index was Rs.10 and at 1992 price index Rs.100.

Five days old pupae can be stored at 10°C for three weeks without any adverse effect on the fecundity of resulting adults.

2.2.2 Field Release

Before releasing in the field in endemic areas, moderate to severely infested plants are marked. The plant stem trunks are ringed one foot away with a band of 5% Diazinon granules 24 hrs before the release of the beetles; this stops the patrolling of ants on the trunk at least for 3 days. On citrus, 10 beetles are released per infested plant but on other crops the releases are calculated based on degree of infestation and density of crop canopy. All precautions should be taken to avoid scarcity of food for the grubs to avoid cannibalism. All the pumpkins showing signs of rotting are to be incinerated.

2.3 Production Procedure for Aphidophagous Coccinellids

Aphidophagous coccinellids should be multiplied on their natural host *viz.*, aphids. However, adult aphidophagous coccinellids can feed and produce viable eggs on *Ferrisia virgata*. Hence to economise use of aphid hosts (which are relatively more difficult to multiply than *F. virgata*), adults of coccinellids may be reared on *F. virgata* and their larval stages are multiplied on *A. craccivora*. For better egg production, freshly emerged beetles are pre-fed with aphids for about ten days before releasing on *F. virgata* infested pumpkins.

2.3.1 Cheilomenes sexmaculata (Fabricius)

C. sexmaculata is a very important, polyphagous predator of aphids and other soft bodied insects. It has been recorded in most crop ecosystems, particularly where aphids are serious pests. It has been produced in the laboratory and used for the suppression of *A. craccivora* on groundnut. *C. sexmaculata* is also used in cotton, citrus and sunflower ecosystems, etc.

2.3.2 Production Procedure

The initial culture is started by collecting unparasitised pupae of *C. sexmaculata* from the field. The parasitised pupae look dark black to brown and are brittle to touch, whereas, healthy puape are bright yellow coloured with black markings and are soft to touch. The freshly emerged adults are fed with *A. craccivora* for ten days. Ten days old, pre-fed beetles are released on pumpkins bearing colonies of 25 days old *F. virgata*. Dry leaves of *Bauhinia purpurea* are arranged on the periphery of the pumpkins in a slanting position. Any other plant's leaves can also be used as an oviposition substrate. The leaves should be broad and should remain stiff after drying. These leaves act as oviposition substrate and also helps the beetles to climb over the pumpkin. Honey (50%) soaked in absorbent cotton swab is provided as an additional carbohydrate source. The beetles are allowed to feed on *F. virgata* in the same cage for 25 days. The pumpkin is replaced with another infested pumpkin if necessary. Similarly, dead adults are removed regularly and honey swabs are changed on alternate days. Dead adults are replaced with fresh ones to ensure regular supply of eggs. The eggs deposited on the leaves are collected and used for further multiplication. Eggs thus collected are kept separately in glass vials, as the grubs are highly cannibalistic. The grubs are released on cowpea seedlings infested with *A. craccivora*, grown either in plastic cups or glass jars. Twenty-five grubs can be reared for 8-9 days in one jar containing more than 6500 aphids. Four jars are required for rearing 100 grubs up to the pupal stage. The adults emerging in 5-6 days from pupae are used either for further multiplication or for field release. The adult beetles survive for 50-70 days on mealybugs. A regular decline in fecundity is observed 3-4 weeks after emergence and the beetles stop laying eggs during the last week of their life. Hence, the beetles should be discarded after the fourth week as they become less productive and rearing becomes uneconomical. Instead of pumpkins, potato sprouts in glass jars can also be used for the multiplication of this coccinellid. Dried leaves are kept at different places in the jar. Egg removal becomes difficult in this procedure. The chances of eggs getting damaged while removing them from the jars are also high. A recent study indicated that the potato plants in polyurethane cups could be used for rearing *C. sexmaculata*. The infested potato plants are kept in the cage and beetles are released. Provision of additional oviposition substrate is not essential in this procedure as the beetles prefer to deposit eggs on the potato leaves.

Coccinella septempunctata can also be reared by following a similar procedure. Because of its higher feeding potential, only fifteen grubs can be reared in one jar containing 100 infested cowpea seedlings. Seven jars are required for rearing 100 grubs upto the pupal stage.

2.4 Production protocols for Anthocorid predators (Ballal and Gupta, 2007)

Anthocorid predators are recognised as potential biocontrol agents in different parts of the world. Natural populations of these predators have been successful in maintaining the pest populations at low levels. In countries like France, UK, Netherlands, Germany, etc anthocorid predators have been multiplied and released in the green houses and fields for management of insect pests. They feed on small lepidopterous larvae, small coleopterous grubs, psocids, mites, thrips & aphids and are represented in all the zoogeographical regions of the world.

Attempts have been made to mass rear anthocorids in other countries - *Anthocoris gallarum-ulmi* (DeG.) on aphid species like *Myzus persicae* Sulzer, *Acyrthosiphon pisum* (Harris) and *Aphis fabae* Scopoli, *Orius insidiosus* Say on eggs of *Dacus dorsalis* Hend. , *Xylocoris flavipes* (Reuter) on larvae of *Tribolium confusum* Duv. or *Tribolium castaneum* (Hbst.); *Xylocoris sordidus* (Reut.) on *T. confusum*. In India, very few attempts were made to rear the anthocorid predators.

At the Project Directorate of Biological Control, a method has been standardized for mass rearing *Cardiastethus exiguus* Poppius, a potential anthocorid predator of *Opisina arenosella* Walker on eggs of alternate host, *Corcyra cephalonica* Stainton. Utilizing 1 cc of *C. cephalonica* eggs, 450 *C. exiguus* adults could be produced (Ballal *et al.*, 2003a). Investigations have also been conducted on another anthocorid predator - *Blaptostethus pallescens* Poppius. The biology and feeding potential of this anthocorid has been studied (Ballal *et al.*, 2003b). Approximately 100 UV irradiated *C. cephalonica* eggs are required for rearing one nymph throughout its nymphal period of about 16 days and 630 eggs for an adult throughout its longevity of 38 to 78 days. Utilising 9cc of *C. cephalonica* eggs, 1000 *B. pallescens* could be produced. Initial trials have indicated the potential of this predator as a bio-agent for the management of aphids, red spider mites and stored grain pests.

3. Field evaluation of *Cardiastethus exiguus* on *Opisina arenosella*

Field trials at Kerala and Karnataka have indicated that *C. exiguus* is a potential predator against *O. arenosella.* At Channapatna, monthly observations were recorded on pest population per leaflet and *Cardiastethus* population per leaflet. Six field releases of the anthocorid were made. There was a significant reduction in the larval population per leaflet in the treated trees. During the last two observations, *O. arenosella* larvae were in late instar and the anthocorid count was also less. This study indicated that the anthocorids are effective against the younger stages of *O. arenosella* and hence the releases should concentrate on those stages. Field releases conducted at Kerala indicated that there was a significant reduction in the population of *O. arenosella* by releasing *C. exiguus* thrice @ 50 adults/nymphs per tree per release (when the pest is in the egg or first instar stage), in comparison to control (Lyla *et al.*, 2006).

Microbials

Bacillus Thuringiensis

Bacillus thuringiensis (Berliner) (*Bt*) is the most widely used microbial insecticide in the world. *Bt* is claimed to be increasingly cost effective and hence is more competitive with chemical insecticides. Several *Bt* strains have been isolated in India from soil or phylloplane of crops. Various *Bt* products have been evaluated in the field at doses varying from 1-2 kg/ha and some pests of field crops like *Helicoverpa armigera, Cnaphalocrocis medinalis, Earias vittella* and *Pectinophora gossypiella* have been controlled successfully in the field.

Bt is available commercially in the form of liquid or powder based formulations. Based on the spore/crystal concentration the formulation is usually used @ 0.2%. However manufacturers specify the dosage. The spray mixture is used to cover both the lower and upper surface of foliage and sprayed preferably in the evenings. Bt formulations are used against various pests as indicated below.

- Multiplication of *Bt*
  Shake flask cultures can be produced on a rotary shaker at 200 rpm and 30°C. Spore-crystal proteins complexes are recovered after 72 hr.

- Batch culture experiments can be carried out in a fermentor. The impeller speed is maintained at 350 rpm, airflow at 5-7 liters/min at 0.5 bar. The spore-crystal proteins complexes are recovered by continuous flow centrifugation (Heraeus) and then eventually formulated as wettable powder by drying or the ferment is formulated directly for bioassays.
- The number of viable cells and free-spores can be counted separately using a Neubauer bacterial counting chamber under phase contrast light microscope.

4. Entomopathogenic Viruses

Nuclear polyhedrosis viruses have enormous potential as viral biopesticides. Research over the last three decades on *Ha*NPV has focused on the natural occurrence of the virus, virus-host relationship, mass production and formulation, field-testing and safety evaluation against several non-target species. Being an obligate pathogen, the virus can be mass produced only in *H.armigera* larvae. Obtaining high yield/larva is critical to economic production of the virus. A successful technique of mass production of *H. armigera* for *in vivo* production of the virus has also been developed.

As baculoviruses are obligate pathogens, availability of their natural live host insect is the foremost requirement for baculovirus mass production. For large-scale production of nuclearpolyhedrovirus of H. armigera and S. litura there must be a large scale culture of respective host larvae. For maintaining the large-scale culture of host insects, host plants such as castor, cauliflower, cotton leaf etc are necessary. Rearing on host plants is labour intensive and can cause cross contamination. To overcome these difficulties artificial diets are used which facilitate large-scale host rearing throughout the year under controlled conditions without depending on the host and season, resulting in an assured supply of adequate and healthy host material.

4.1 Mass Multiplication of *H. armigera*

For the multiplication of natural enemies like parasitoids or NPV, it is important to mass multiply H. armigera in large numbers.

4.2 Mass Production of Virus

When the larvae reach early fifth instar, 90% of the larvae are used for NPV production, whereas remaining 10% is retained for continuing the healthy culture. Artificial diet is prepared as mentioned earligr'without formalin. 10"ul og virus suspension (5 x 10^7 POB/ml for *H. armigera* and 5 x 10^8 POB/ml for *Spodoptera litura*) is dispensed and spread onto the artificial diet. Early fifth instar larvae are released individually in these vials. All the vials are incubated at 25^0C. After 6-7 days the larval mortality is observed. All the dead larvae are collected, macerated, semipurified (by differential centrifugation) and enumerated using double ruled Neubauer Haemocytometer. And the virus thus collected is preserved in the refrigerator till further use.

While crude NPV preparations were found to be effective in the control of H. armigera, development of stable basic formulations of the virus was considered critical for commercial scale production. In India, wettable powder formulation of HaNPV was developed and was found to be effective in field tests (Rabindra et al., 1990). The virus at a concentration of 1.5x1012 POB/ha was found to control the population of H.armigera and increase the yield of chickpea, pigeon pea, sunflower, tomato, groundnut and Egyptian clover. On cotton, a higher dose of 3x1012 POB/ha was required. The virus has been used in combination with other components of pest management like Bt, chemical insecticides, azadirachtin, neem seed kernel extract, Chrysoperla carnea and bird perches.

Control of *S. litura* has been studied on banana, black gram, castor, cauliflower, cotton and tobacco using NPV alone. Granulosis viruses have been reported from several insect pests in India. A granulosis virus infection was reported on *Chilo infuscatellus.* In several field trials, *CiGV* was found to be effective in controlling the shoot borer in sugarcane field. Spraying the GV in potato fields in Maharashtra @ 5.5x10^{12} OB/ha + endosulfan (0.035%) three times at 10 days interval gave excellent protection to the tubers from Potato tuber moth. A highly virulent GV was isolated from the larvae of the diamondback *P. xylostella.* Investigations were made on the mass production, formulation and field evaluation of the *Px*GV against the DBM on cauliflower.

4.3 Quality Control

- 1. Viral Unit (NPV) - 1 x10^9 POB/ml or g (for 9(3) registration) 1 x 10^8 POB/ml or g (for 9(3b) registration)
- Contaminants such as *Salmonella, Shigella* or *Vibrio* should be absent. Other microbial contaminants should not exceed 1 x 10^4 Count/ml or g.

- An undertaking should be submitted that the strain is indigenous, naturally occurring not exotic and not genetically modified.

5. Entomopathogenic Fungi

Entomopathogenic fungi are reported to cause epizootics in insect pest population indicating great potential for pest management. Species that have been most intensively investigated as mycoinsecticides in the crop pest control include *Beauveria bassiana, B.brongniartii, Metarhizium anisopliae, M. flavoviride, Verticillium lecanii, Hirsutella thompsonii, Nomuraea releyi, Paecilomyces farinosus, P. fumosoroseus, Entomophthora* sp., *Fusarium* spp., *Aspergillus* sp. *B.bassiana* and *M.anisopliae* have been used on a large scale for the control several crop pests over a number of years, while others have been attempted in recent times.

5.1 Mass Production of Entomopathogenic Fungi

Most of the entomopathogenic fungi can be grown on a variety of solid substrates or in liquid media. Sorghum grain, rice, rice flakes, puffed rice, wheat, wheat flakes, maize, cowpea grains, millets, rice bran, wheat bran, groundnut hull meal, bengal gram husk, coffee husk, potato, carrot etc. are some of the solid substrates used for mass production of *B. bassiana, M. anisopliae, V.lecanii, P.farinosus* and *N.rileyi*. These solid substrates are moistened, sterilized and inoculated with the fungus in bottles/polypropylene bags and incubated at optimum temperature and humidity for a period of 10- 20 days. Then the viable propagules are harvested directly in water containing wetting agent and vegetable oil by repeated washing or centrifugation and used as foliar sprays. Alternatively, talc based formulations can be prepared by homogenizing the substrate along with fungal growth into a fine powder and mixing with inert materials like, talc (1:1 proportion) and used for foliar sprays The spore concentration in the formulation should be more than 1X 10^8spores/g/ml

In liquid cultures, these fungi can be grown in static liquid cultures or shake cultures or in fermentors using cheap and inexpensive liquid media like molasses, carrot extract, potato extract or synthetic media. The infective propagules can be harvested by filtration/centrifugation and mixed with water containing wetting agent and vegetable oil to get spore suspension with required spore concentration. In case of certain fungi like, *N.rileyi, B.bassiana* and *M.anisopliae* submerged cultures give rise to blastospores or a mixture of blastospores and conidia. The blastospores are highly sensitive to high temperatures and other environmental stresses and their viability is lost quickly. To overcome this difficulty, diphasic fermentation technique is used for maximum production of aerial conidia. In this method, the fungus is allowed to grow in fermentor upto the end up the log phase for maximum production of mycelial biomass, then it is subsequently transferred to nutritious (grains) or inert substrates (talc, clay granules) for production of aerial conidia in the form of natural inocula.

6. Entomopathogenic Nematodes (EPN) (Hussaini, 2007)

Entompathogenic nematodes popularly called as EPNs offer great scope for the management of many lepidopterans insect pests. In India the work on steinernematids started in the sixties with the use of DD-136 for the control of pests of rice, sugarcane and apple. Recent field studies confirmed the efficacy of *S. carpocapsae* against the brinjal shoot and fruit borer *Leucinodes orbonalis.* At a dose of 5 billion ijs/ha, the yield of fruits could be increased by 20 percent.

6.1 Mass Production

Mass production is a key issue to EPN commercialization. Identification of potential isolate, characterization of symbiotic bacteria and suitable media are important factors. Axenic culture process for *S. glaseri* and *S. carpocapsae* was developed in 1940 followed by an inexpensive medium for *C. elegans* with high yields. A variety of substances; Potato mash, ground veal pulp, peptone-glucose agar and pork kidney, homogenized animal tissue, dog food, chicken offal homogenate. Modified dog biscuit, egg yolk and soy flour and cholesterol medium were suitable for indigenous isolates.

In the *in vivo* process, an insect serves as a bioreactor; in the *in vitro* process, artificial media are used. For testing in field experiments larger quantities of nematodes are necessary, which are preferably produced near the testing facilities in order to maintain nematode quality. Although *H. indica* is more resistant to heat stress than *H. bacteriophora,* a decrease in the namatode's control potential due to high temperature during long distance transportation can severely decrease nematode survival and control potential.

EPNs possess specific biological and ecological features, which make their use in biological control exceptionally safe. Using the *in vivo* process, yields between 0.5 x 10^5 and 4 x 10^5 IJs/larva, depending on the nematode species, have been obtained in other countries. During the past few years, a distinct cottage industry has emerged that utilizes the *in vivo* process for nematode mass production for sale, especially in the home lawn and garden markets. However, the *in vivo* process lacks any economy of scale; the labour, equipment, and material (insect) costs increase as a linear function of production capacity. Perhaps even more important is the lack of improved quality while increasing scale. The *in vivo* nematode production is increasingly sensitive to biological variations and catastrophes as scale increases. In a scale-up model, Friedman (1990) reported that the solid-culture method is economically feasible up to a production level of approx 10×10^{12} nematodes/mo. Labour costs increase significantly for nematode production beyond this level, making a less expensive method of large-scale production a necessity. Following recovery from production substrates, the nematodes can be either stored in bulk for extended periods or formulated immediately. When the nematodes are to be stored as bulk, nematode concentration, temperature, aeration, and contamination control are important considerations for the maintenance of high viability and quality. Differences in storage stability among nematode species can be attributed to their thermal and behavioral adaptations. Each nematode species has a well-defined thermal niche (Grewal, *et al.*, 1994) and an optimum temperature for the longest storage stability. *S.feltiae* stores better at 5°C, and *S. scapterisci*. *S. riobrave* and *H. bacteriophora* are more stable at 10°C. *S. carpo*capsae and *S. scapterisci* that adopt a quiescent posture during storage generally store better than the more active species, such as *H. bacteriophora* and *S. riobrave*. The latter species also tend to be more prone to bacterial contamination during storage. Antimicrobial agents may be used to suppress contamination, and the choice of the antimicrobial should be based on its safety to nematodes and symbiotic bacteria.

7. Issues Involved in Adoption Biocontrol Strategy

The major issues involved in the adoption of biological control are - selection of best species and strains, development of cost-effective methods for mass rearing, testing quality, development of effective methods for storage and shipment, creation of formulations of pathogens to protect and deliver the agent to the target, development of methods for release/application, field evaluation, critical economic factors affecting product development and markets, demonstration of safety of products to man and the environment, feasibility of continuity of the system, etc.

Production and storage of entomopathogenic fungi

Pathogen	Production method	Media	Form of Inoculum
Aschersonia aleyrodis	Surface or submerged culture, Semi-solid or Diphasic culture	PDA or Chopped millets	Dry mycelium, Blastospores, Conidia
Beauveria bassiana	Surface or submerged culture, Semi-solid or Diphasic culture	Routine agar or liquid media	Dry mycelium, Blastospores, Conidia
B.brongniartii	Surface or submerged culture, Semi-solid or Diphasic culture	Routine agar or liquid media	Dry mycelium, Blastospores, Conidia
Entomophaga spp.	Surface or submerged culture, Live host	Sabouraud dextrose egg yolk milk agar (SEMA)	Conidia
Erynia neoaphidis	Surface or submerged culture, Live host Semi-solid or Diphasic culture	SEMA	Dry mycelium, Conidia
Hirsutella thompsonii	Surface or submerged culture, Semi-solid or Diphasic culture	Routine agar or liquid media	Submerged Conidia, Blastospores
Metarhizium anisopliae	Surface or submerged culture, Semi-solid or Diphasic culture	Routine agar or liquid media	Dry mycelium, Blastospores, Conidia
M. flavoviride	Surface or submerged culture, Semi-solid or Diphasic culture	Routine agar or liquid media	Dry mycelium, Blastospores, Conidia
Nomuraea rileyi	Surface, Semi-solid or Diphasic culture	Sabouraud maltose agar with yeast extract (SMAY)	Dry mycelium, Blastospores, Conidia
Paecilomyces farinosus	Surface or submerged culture, Semi-solid or Diphasic culture	Routine agar or liquid media	Dry mycelium, Blastospores, Conidia
Verticillium lecanii	Surface or submerged culture, Semi-solid or Diphasic culture	Routine agar or liquid media	Dry mycelium, Blastospores, Conidia
Zoophthora radicans	Surface or submerged culture, Live host Semi-solid or Diphasic culture	SEMA	Dry mycelium, Conidia

7.1 Shipment

Shipment is generally done by quickest means. Care should be taken about the prevailing temperature, humidity and time taken in the transit and the container should be escape proof. Food for bioagents, including honey, split raisins, sugar cubes, honey-agar, and sugar candy may be provided while shipping the live adult biocontrol agents. All instructions for storage and delivery should be written or printed in bold letters in different colours for easy understanding of the receiver. Proper packaging of the bio-agents to be shipped is of utmost importance, especially in the case of parasitoids and predators. Packaging techniques for Tricho-cards and *Telenomus remus* parasitised egg cards have been standardized.

8. Future Thrust Areas in the Field of R&D for Promotion of Biological Control

- Determine threshold levels for different pests with reference to use of bio-agents
- Identification and development of improved strains of biocontrol agents
- Standardisation of low-cost production technology for the potential bio-agents
- Methods to be devised for scaling up, mechanizing and improving the efficiency of the bio-agents.
- Methods to increase shelf life of bio-agents
- Cost-effective improved and stable formulations of microbials
- Development of products with enhanced shelf life
- Development of UV-tolerant strains of microbials
- Development of formulations with enhanced persistence on host plants, particularly cotton
- Improved and mechanized application technology
- Establishment of quality control laboratories
- Refine of techniques to assess the impact of natural enemies
- Training to be strengthened for government and private units on production, evaluation and quality control aspects of bio-agents

- Industry-Institute joint collaboration for working on the above thrust areas
- Promotion of organic farming as an integral part of organic farming.

❑❑❑

13

Improved Mass Production Technologies of Major Biocontrol Agents for Efficient Pest Management

VIJAYAKUMAR, N.
Entomologist, Bio Control Laboratory, Perunthalaivar Kamaraj Krishi Vigyan Kendra (PKKVK), Kurumbapet, Puducherry - 605 009, India.

In India, the first mass production unit for large scale production and use of *Trichogrammatids* for the control of Sugarcane Tissue borers was started at Mandya during the year 1935 and Rangaswami *et al.* (1968) was the first to report, feasibility of microbial control of Crop pests like *Amsacta albistriga* (Walker) *Helicoverpa armigera* (Huebner), *Earias fabia* (Stoll.) *Nephantis serinopa* (Meyrick) with fungi like *Metarrhizium* sp., *Beauveria* sp. and *Aspergillus* sp. under field conditions. Since then large number of scientific acumen associated in the production of biocontrol agents for the biological suppression of crop pests and diseases.

1. Facilities Required for Production of Biocontrol Agents

The mass production facilities may vary depending on the type and number of Bioagents to be produced. The basic amenities such as well lighted and ventilated rooms of different sizes preferably with low roofing and window facing north fitted with insect escape proof mesh are to be constructed in the groves of shade trees. Modern facilities like insulated walls, low insulated ceilings, large windows fitted with glass shutters, sufficient lighting arrangement and central or split type or window type Air Conditioners are to be provided while planning new buildings to house mass production units besides to have methods to maintain cleanliness and to service the equipments at affordable costs.

Rooming facilities for storage of Diet ingredients, unused cages and other materials, Diet preparation, Egg or larval implantation, Insect development, Adult emergence are absolutely essential and the size of the rooms depends on the organism to be produced and the quantity of the end product requirement. One of the prerequisite for mass production rooms should be in general sanitation in order to prevent the contamination due to the entry of unwanted organisms.

As the environment needs the different ambience for various bioagents, the specific room or rooms may have the right ambience of temperature, humidity and photoperiod suitable to particular organism and in general and the temperature is maintained at about 25° to 28°C and relative humidity 55 % to 65 % .The facilities includes racks, tables, chair, and stools of suitable sizes. The working table and racks are fitted with castor wheels so that routine cleaning or shifting within the room or from one room to another room becomes easy. The racks should have arrangements for adjusting the partition depending upon the heights of the cages to be used. Besides this, Microscope with Phase contrast field, Laminar flow and High efficient particulate air filter, standard furnace filters, Air lock system, Autoclave, Hot Air Oven, Table lamps, Walk-in-incubators, Ultra Violet lamps, Formaldehyde Gas fumigation chambers, Distillation unit, Refrigerator, Deep Freezers and other miscellaneous equipments for day today mundane works are required.

2. Rearing Requirements for Production of Host Insect and Biocontrol Agents

Different types and sizes of rearing containers/cages are used for insect rearing with transparent material on one side or top for easy viewing. Basically the rearing containers are made of materials which are non toxic, cheap, optimum sized to permit mating and host searching, and amenable to easy cleaning and disinfection and are provided with insect escape proof net or cloth. Most commonly used cages are wooden cages, which are now being replaced by plastic or aluminium cages. The cages are covered by cloths on all the sides except the front door which is fitted with clear acrylic sheet for viewing. Sleeves are provided on one or two sides or on top for easy handling of the materials inside.

In order to undergo individual insect rearing to prevent cannibalism, tubes (plastic or glass), empty injection vials covered with cotton plug or a piece of cloth fastened with rubber bands are used. Very often clear transparent containers are modified to serve as cages by cutting windows on the sides and top cover is covered with insect escape proof mesh.

Pumpkins are kept in cage for multiplication of scale insects, mealybug parasitoids and predators for a long time. Cages 0.3 sq m with cloth fixed on all sides and two sleeves provided on sides or one on top are used. The bottom or base of theses cages are made of the same material as the frames.

Some of the parasitoids and even their hosts require more humidity, which is provided by keeping moist sponge or moss in the cages and spraying water over the cloth of the cages.

Moist saw dust, sterilized sand or even the rooted canes or potted plants of paddy are kept in aluminium tray filled with water which is replenished every alternate day for multiplication of Natural Enemies of sugarcane and paddy pests in high humidity conditions.

Eggs parasitoids of leafhoppers and other pests, adults of which are strongly attracted towards light, are reared in corrugated boxes lined inside with black cloth or paper . The emerging adults are collected in glass tubes fixed on top or side of such boxes. Even ordinary cages in use can be covered with black cloth on all the sides and glass tubes fixed either on front or top to serve the same purpose.

Potted plants infested with host material are kept in large sized cages measuring ½ x ½ x 1 mt, or the plants are grown in the field and covered with field cages measuring 3 x 3 x 2 mts with fibre glass top can be used for multiplying Nature and the NE/host population is regulated constantly by continuous monitoring and collection Natural Enemies.

Adults are fed on dried, boiled or soaked raisins, sugar cubes, 5-10% sugar solution, pure honey, 25 or 50% honey, protinex, fructose, castor pollen, honey-protinex, agar-sugar candy, etc.

The immature stages of Natural Enemies are multiplied on factitious hosts on artificial or natural diets or on target host. Semi-synthetic diets have been evolved for a number of host insects which have facilitated multiplication of host specific natural enemies. In all mass multiplication programmes, performance and quality of natural enemies produced must be tested so that Natural Enemies of high quality are released in field.

3. Selection of Biocontrol Agents

According to Wage (1997), an ideal Bio control agent that the farmer would like to have is the one which could not only kill the pest but also provide long-lasting and widespread control by its own reproduction. It is mandatory that the scientific name and the strain of Bioagents to be mass produced should be known and it is also essential to ascertain the purity of the culture before the start of selection of Bioagents for mass multiplication. The Bioagents selected should be an effective species or strain, amenable to mass production, capable of developing on resistant / tolerant cultivars, compatible with other crop production inputs with high reproductive potentials suitable for particular climatic niche, preferably tolerant to most common pesticides, with high host searching capacity suitable for seasonal inoculative and inundative use and also free from patent disputes.

4. Availability of Production Systems of Biocontrol Agents

In order to have continuous cultures of host insects, the host insects could be reared on natural hosts (eg. Aphids), on factitious hosts (e.g. Scales and Mealy bugs on pumpkins) and on semi - synthetic diets (e.g. lepidopterous larvae). Some of the most commonly used host insects are *Corcyra cephalonica* (production of trichogrammatids & chrysopids), *Sitotroga cerealella*, etc.

Improved techniques have been standaridized for multiplication of 10 dozen bacteria. Microsporidia (*Nosema* sp.), bacteria (*Paenibacillus popilliae*, *P. lentimorbus*), Nuclearpolyhedroviruses (NPVs), Granulosisviruses, parasitoids (hymenopterans, dipterans, etc.), predators (coccinellids, syrphids, etc.) and nematodes (*Steinernema* spp. *Heterorhabditis* spp.) could be produced under *in vivo* conditions. The antagonists of plant diseases and phytonematodes could be multiplied on locally available cheap materials such as grains, molasses and other industrial wastes.

MASS PRODUCTION OF HOST INSECT *CORCYRA CEPHALONICA*

2.5 kg cleaned insecticide-free, fresh Bajra / Sorghum / Maize / Paddy grains, sterilized at 100°C for 30 min. and put into culturing plastic square trays (One Set used for 100 Days)

↓

Incorporating powdered Groundnut kernel weighing 100 g as a source of protein

↓

Adding yeast tablets (10 Nos or 5 g) as nutritional supplement + 0.05g Streptomycin sulphate as Antibiotic. Supplementing 0.01 g Sulphur as Acaricide

↓

Mixing one cc afresh and viable eggs of *C*orcyra in each tray

Tray covered with muslin cloth. Maintenance of culture at 26°C

↓

After 40-45 days, emergence of adult moth commences

Regular collection of adult moths bot by manually and mechanically using Vacum cleaner

↓

Allow adult moths in Fecundity cage used for 3 Days supplementing with food swab impregnating Honey 15%, Water and Vitamins E (2 cap.)

↓

Daily collection of *Corcyra* eggs

↓

Pass the eggs through 15, 30 and 40 mesh sieves and run over a slop of paper to eliminate dust particles or Separation of scales admixtured in eggs Mechanically by Motorised Egg Separator

↓

Harvest of fresh eggs

↓

Usage of sieved eggs for "*Parasitization*"

↓

Storage of Eggs at 10°C upto 7 days

MASS MULTIPLICATION OF *TRICHOGRAMMA* PARASITOIDS

Egg parasitoid – *Trichogramma species*

↓

Afreshly collected Corcyra eggs are subjected to UV treatment using 30 watt UV tube light for 45 min. at a distance of 2'

↓

Pasting UV treated eggs on to the printed "*Tricho*" cards measuring size 15cm x 10cm comprising six pieces (12 x 3 cm each)

↓

Introduction of single egg card inside the polythene balloon along with nucleus culture of "*Trichogramma*" strain maintaining in ratio of 1 Female to 30 Eggs for effective parasitization

↓

Parasitization takes place within a week time or ten days

↓

About 12,000 *Trichogramma* adults emerge out from a single card

↓

Tiny wasp, life cycle 8-12 Days

↓

Biology of *Trichogramma*

Egg Period	:	1 Day	Pupal Period	:	4-5 Days
Larval Period	:	2-3 Days	Adult Longevity	:	6-8 Dyas

↓

General maintenance of various species *Trichogramma japonicum* and *Trichogramma chilonis*

↓

Quality control, check once in a month

↓

Shelf life 2-3 days

MASS PRODUCTION OF HOST INSECT *SPODOPTERA LITURA*

In vivo collection of Egg masses of *Spodoptera litura.* Transferring neonated larvae into fresh castor leaves or Semi Synthetic diet upto fifth instar

↓

Larvae allowed to Pupate

↓

Surface sterilization of cocoon with 0.1% Sodium Hypochlorite and washed in running tap water

↓

Drying and place them on filter paper toweling moistened sponge in a container

↓

Emergence of Adults after seven days

↓

Transferring emerged adults for egg laying into plastic container supplemented with 50 % water + Honey swab as food

↓

Cycle continues and supplying third instar larvae for NPV production (5 % for recycling and 95 % for NPV production)

↓

Biology of *Spodoptera litura*

Egg Period	:	3-5 Days	Pupal Period	:	9-12 Days
Larval Period	:	21-28 Days	Adult Longevity	:	7-8 Days

MASS PRODUCTION OF NUCLEO POLYHEDRO VIRUS FOR *SPODOPTERA LITURA*

Third instar larvae (7-9 days) are subjected to starvation for 8 h.

↓

Preparation of virus suspension containing 1 x 10^9 POB / ml in 100 ml of distilled water containing 0.1 per cent Teepol

↓

Treating clean Castor leaves with virus suspension by dipping and drying under shade or smearing viral suspension onto the artificial diet cake using glassrod

↓

Allowing the caterpillars to feed on the treated leaves for 2 days and subsequently on untreated leaves or Release the larvae into the plastic bottles containing Semi-Synthetic diet impregnated with NPV

↓

Collecting the dead larvae due to NPV infection in distilled water and allowing to stand for a week for Putrification

↓

Mascerating the larvae in Distilled water or in 0.1 % SDS using blender for homogenization of suspension

↓

Filter the homogenized solution through muslin cloth

↓

Purify Polyhedra by Centrifuge the filtrate for 5 min. at 500 rpm

↓

Collect supernatant and again centrifuge for 30 min. at 10,000 rpm Supernatant is discarded

↓

The pellet is resuspended in small volume of water or 0.1 % SDS

↓

Virus particles as polyhedra settle down as white layer

↓

Collect sediment with distilled water and keep in glass container / bottle as Primary Suspension (pure *Sl* NPV)

↓

Quality Testing 1 LE = 1 x 10^9 POB / ml by Haemocytometer

↓

Use Sodium Alginate 2 g, Calcium Chloride 1 g, Borax 1 g Polyacrylamide 0.5 g and Acetone 1 ml Precipitation

↓

Formulation of NPV Biopesticide

↓

Store the pure *Sl* NPV in refrigerator in amber coloured bottle at 5°C for further use

↓

Shelf life 6 months to 1 year under refrigerated conditions at 5°C and years together under deep freezer at -18°C

MASS PRODUCTION OF HOST INSECT *HELICOVERPA ARMIGERA*

In vivo collection of Egg masses of *Helicoverpa armigera*

↓

Transferring neonated larvae into soaked chickpea or Semi Synthetic diet upto fifth instar in individual vials or multicellular Acrylic cages or Acrylic louver cages

↓

Larvae allowed to Pupate

↓

Surface sterilization of cocoon with 0.1% Sodium Hypochlorite and washed in running tap water

↓

Drying and place them on filter paper toweling moistened sponge in a container

↓

Emergence of Adults after seven days

↓

Transferring emerged adults for egg laying into plastic container supplemented with water + Honey swab as food

↓

Cycle continues and supplying third instar larvae for NPV production (5 % for recycling and 95 % for NPV production)

↓

Biology of *Spodoptera litura*

Egg Period	:	4-5 Days	Pupal Period	:	10-12 Days
Larval Period	:	17-20 Days	Adult Longevity	:	9-10 Days

MASS PRODUCTION OF NUCLEO POLYHEDRO VIRUS FOR *HELICOVERPA ARMIGERA*

Fourth instar larvae are subjected to starvation for 8 h.

↓

Preparation of virus suspension containing 1 x 10^9 POBs / ml to 100 ml of distilled water containing 0.1 per cent Teenopol

↓

Heads of these larvae are dipped in NPV suspension (or) pour a drop of NPV suspension into Head capsule of larvae using Micropipette

↓

Larvae are reared individually in vials / cells of either Multicellular Acrylic Cages or Acrylic Louver Cages with soaked chickpea (or) Semi-synthetic diet

↓

Collecting the dead larvae due to NPV infection in Distilled water and allowing to stand for a week for Putrification

↓

Mascerating the larvae in Distilled water or in 0.1 % SDS using blender for homogenization of suspension

↓

Filter the homogenized solution through muslin cloth

↓

Purify Polyhedra by Centrifuge the filtrate for 5 min. at 500 rpm

↓

Collect supernatant and again centrifuge for 30 min. at 10,000 rpm Supernatant is discarded

↓

The pellet is resuspended in small volume of water or 0.1 % SDS

↓

Virus particles as polyhedra settle down as white layer

↓

Collect sediment with distilled water and keep in glass container / bottle as Primary Suspension (pure *Ha* NPV)

↓

Quality Testing 1 LE = 1 x 10^9 POB / ml by Haemo cytometer

↓

Use Sodium Alginate, Calcium Chloride, Acetone Precipitation

↓

Formulation of NPV Biopesticide

↓

Store the pure *Ha* NPV in refrigerator in amber coloured bottle at 5°C for further use

↓

Shelf life 6 months to 1 year under refrigerated conditions at 5°C and years together under deep freezer at -20°C

MASS PRODUCTION OF GRANULOSIS VIRUS FOR SUGARCANE SHOOT BORER *CHILO INFUSCATELLUS*

Fourth instar larvae are subjected to starvation for 8 h

↓

Preparation of virus suspension containing 1 x 10^9 POBs / ml to100 ml Distilled water containing 0.1% Teenopol

↓

Head capsule larvae are fed with a drop of Granulosis Virus suspension through a pin head or by dipping the head of the larva into GV. Suspension

↓

Virus fed larvae are reared on sugarcane shoot bits @ 3 per plastic box measuring 7 cm x 7 cm x 6.4 cm

↓

Collecting the diseased larvae due to Granulosis Virus Suspension in 0.1 % SDS and allowing to stand for a week for putrification

↓

Macerating the larvae in Distilled water or in 0.1 % SDS using blender for homogenization of suspension

↓

Filter the homogenized solution through muslin cloth

↓

Feed fresh beetles on *Aphis craccivora* for 10 days

↓

Release 10 day old pre fed (20 pairs beetles on *Ferrisia virigata* infersted pumpkin with 25 day old mealy bugs) and dry leaves provided as oviposition substrate in each cage

↓

Allow beetles to feed and ovipostate on dry leaves in the same cage for 25 days

↓

Replace the pumpkin if necessary

↓

Collection of eggs alongwith dry leaves daily for 25 days

↓

Collection of grubs and adults within 5-6 days

↓

Biology of *Coccinella septumpunctata*

Egg Period	:	3-4 Days	Pupal Period	:	5-6 Days
Larval Period	:	7-9 Days	Adult Longevity	:	40-50 Days
Fecundity	:	200 – 300 Eggs			

MASS PRODUCTION OF *CHEILOMONES SEXMACULATA*

Sow 100 pre-soaked Cowpea seeds in a jar 20x15 cm containing sand

↓

Inoculate sprouts on 4th day of emergence with 100 Apterous females of *Aphis craccivora*

↓

Within a week 6500 *Aphis craccivora* develop in a jar

↓

Rearing the hatched grubs on *Aphis carccivora* in jars

↓

4 jars require for rearing 100 grubs in 7-8 days

↓

Adult emergence

↓

Collection of 100 fresh adults from 4 jars

↓

Feed fresh beetles on *Aphis carccivora* for 10 days

↓

Release 10 day old pre fed (20 pairs beetles on *Ferrisia virgata* infested pumpkin with 25 day old mealy bugs) and dry leaves provided as oviposition substrate in each cage

↓

Allow beetles to feed and ovipostate on dry leaves in the same cage for 25 days

↓

Replace the pumpkin if necessary

↓

Collection of eggs alongwith dry leaves daily for 25 days

↓

Collection of grubs and adults within 5-6 days

↓

Biology of *Cheilomones sexmaculata*

Egg Period	:	3-4 Days	Pupal Period	:	4-5 Days
Larval Period	:	6-8 Days	Adult Longevity	:	50-60 Days
Fecundity	:	200 – 300 Eggs			

MASS PRODUCTION OF *SCYMMUS COCCIVORA*

Select and procure medium sized potato showing initiation of sprout from market

↓

Clean with water, treat with 2% benlate shade dry and keep in moist sand for 7-9 days

↓

Infest each sprout with 5-7 gravid females of *Ferrisia virgata*

↓

Keep infested potato in plastic basin (35 x 18 cm) Cover with black cloth and keep in darkened Room for 25 days or in an incubation for 28°C

↓

Keep 3 1/2 Kg. infested potatoes (2 ½ Kg for first 15 days and 1 Kg for another 15 days Bring 25 days old colony of *Ferrisia virgata* in a ventilated plastic jar (25 x 15 cm)

↓

Release 5 pairs of 5 day old pre-fed and pre-mated *Scymmus coccivora* adults in plastic containers

↓

Allow adults to feed and breed in the same jar and add infested potatoes as and when needed

↓

Collection of 275-300 adults in a month time from 1 jar

↓

Storing of adults at 55°C for 15 days

↓

Biology of *Scymmus coccivora*

Egg Period	:	4-5 Days	Pupal Period	:	5-6 Days
Larval Period	:	9-12 Days	Adult Longevity	:	60-65 Days
Fecundity	:	55 / female			

MASS PRODUCTION OF *MALLADA BONINEUSIS*

Collection adults and transfer to cages for rearing with adult (50% honey, protienex mixture & Castor Pollen) for 20 days

↓

Collection of eggs

↓

Larval rearing in groups for 3 days on *Corcyra* eggs

↓

Transfer the larvae to plastic louvers for individual rearing on *Corcyra* till cocoon formation

↓

Collection of cocoons

↓

Adult emergence

↓

Biology of *Mallada bonineusis*

Egg Period	:	3-5 Days	Pupal Period	:	8-9 Days
Larval Period	:	5-6 Days	Adult Longevity	:	40-45 Days
Fecundity	:	300 Eggs			

MASS PRODUCTION OF *CHRYSOPERELLA CARNEA*

Collect adults and transfer to oviposition cage for rearing with adult diet

↓

Egg collection for 25-30 days

↓

Storage of eggs at 10°C for 15 days

↓

Eggs should be shifted to room temperature for hatching

↓

Larval rearing in groups for 3 days on *Corcyra* eggs

↓

Transfer to plastic louvers for individual rearing on *Corcyra* eggs till cocoon formation

↓

Collection of cocoons

↓

Adult emergence

↓

Biology of *Chrysoperella carnea*

Egg Period	:	3-5 Days	Pupal Period	:	5-7 Days
Larval Period	:	8-10 Days	Adult Longevity	:	30-40 Days
Fecundity	:	300-400 Eggs			

❏❏❏

14

Analysis of Natural Enemy Biodiversity in Rice and Cotton Ecosystems and Evaluation of IPM Components

B. ANANDA GANESA RAJA
Lecturer in Entomology, Faculty of Agriculture
Annamalai University, Annamalanagar – 608 002, Tamilnadu

When little or no insecticide is used, tropical irrigated rice fields accommodate a rich arthropod community including different kinds of natural enemies (Greathead, 1979), and in these circumstances, their species richness and abundance may sometimes be greater than those of pests (Heong *et al.*, 1991). In Thailand, food webs comprised over 50 families and many species of natural enemies associated with defoliator, stem borer, hoppers and gall midge pests.

Predator and parasitoid guilds recorded in five different irrigated rice ecosystems in the Philippines included 10 orders and 57 families of predators and 3 orders comprising 40 families of parasitoids, of which 33 species are known natural enemies of pest Homopterous. This abundance and richness, account for biological control potential, contrast to the relative paucity of natural enemies in temperate irrigated rice.

1. Groups of Phytophages

Phytophages (Rice pests)	**Taxonomic groups**
Homoptera	Cicadellidae, Delphacidae, Coccidae, Lophopidae
Heteroptera	Alydidae, Coreidae, Pentatomidae
Thysanoptera	Thripidae
Diptera	Ephydridae, Muscidae, Chironomidae, Cecidomyiidae
Coleoptera	Chrysomelidae, Elatreidae, Curculionidae
Lepidoptera	Hesperiidae, Noctuidae, Pyralidae, Nymphalidae
Isoptera	Termitidae
Orthoptera	Acrididae
Phytophages (non-rice pests/ visitors)	
Lepidoptera	Papillionidae, Lycaenidae, Pieridae, Nymphalidae, Hesperiidae
Hymenoptera	Halictidae, Anthophoridae, Apidae, Megachilidae
Heteroptera	Coreidae, Pyrrhocoridae, Lygaeidae, Anthocoridae, Aphididae
Coleoptera	Chrysomelidae, Curculionidae
Predators:	
Heteroptera	Miridae, Nabidae, Reduvidae, Lygaeidae, Velidae, Mesovelidae, Hydrometridae
Coleoptera	Coccinellidae, Staphylinidae, Carabidae, Tenebrionidae
Orthoptera	Tettigonidae, Gryllidae, Tridactylidae
Hymenoptera	Formicidae, Vespidae, Pompilidae, Sphecidae, Eumenidae
Diptera	Ephydridae, Platystomatidae
Odonata	Libellulidae, Gomphidae, Coenagrionidae, Protoneuridae
Demaptera	Carcinophoridae
Mantodea	Mantidae
Phasmatoidea	Phasmatidae
Neuroptera	Ascalaphidae
Acari	Phytoseiidae

Araneae	Araneidae, Tetragnathidae, Therididae, Linyphiidae, Lycosidae, Salticidae, Thomisidae, Oxyopidae, Clubionidae, Heteropodidae, Corinidae, Philodromidae, Gnaphosidae
Parasitoids	
Hymenoptera	Ichneumonidae, Braconidae, Trichogrammatidae, Eulophidae, Mymaridae, Chalcididae, Pteromalidae, Eupelmidae, Gastreuptiidae, Platygasteridae, Scelionidae, Diapriidae, Bethylidae, Dryinidae, Scoliidae, Metillidae, Tiphiidae
Diptera	Pipunculidae, Tachnidae
Strepsiptera	Halictophagidae, Elenchidae
Scavengers/ Decomposers	
Coleoptera	Meloidae
Diptera	Tabanidae, Sarcophagidae, Anisopodidae, Ceratopogonidae
Blattoidea	Blattoidae
Collembola	Isotomidae, Sminithuridae, Entomobryidae

2. Impact of Natural Enemies on Rice Pests

The potential role of naturally, occurring biological control in tropical irrigated rice has been increasingly appreciated (Nickel, 1964; Yasumatsu and Torii, 1968). Evidence summarized in detail by Greathead (1979) shows, for example, that parasitoids may kill 0-70 of eggs and 0-20 percent of nymphals hoppers. Lim (Lim, 1970)) pointed to the severity of stem borer attack in regions where insecticides were used but not where natural enemies remained unharmed. Parasitoids and predators caused more than 90 per cent of the more than 98 percent mortality from egg to pupa in common stem borer species attacking traditionally grown irrigated rice in Sarawak, and the 'complete control of *Chilo suppressalis* in Hawai was attributed to introduced parasitoids. Against rice bugs, Leptocorisa spp. (Hemiptera: Coreidae), parasitoids caused 11-75 percent egg mortality and predators caused about half of the 90 percent mortality from egg to the final nymphal instar.

3. Implementing Rice IPM

IPM specialists are now advocating radically different message-insecticides are usually not needed in rice-and are focusing on a strong crop in healthy paddy ecosystem. But they have the difficult task of undoing the effect on farmers of decades of exhortations to rely on pesticides. Advocacy of pesticides use, reinforced by extensive agrochemical advertising, led both farmers and policy makers to overestimate crop losses caused by pests, and the effectiveness of insecticides. Unless they understand the benefits of avoiding unnecessary insecticide use, farmers tend to overreact to slight infestations and make routine preventive applications.

Twelve per cent of pesticides sold worldwide are applied to rice crops, and no other single crop accounts for as much pesticide use (Woodburn, 1990). Rice farmers will continue to be the target of massive agrochemical industry marketing and promotion that is supported by financial resources dwarfing those of agricultural extension programmes.

3.1 General IPM Practices for Rice

- Remove and destroy stubbles after harvest
- Trim field bunds
- Provide effective drainage if required
- Avoid use of excessive 'N' fertilizers
- Avoid close planting, especially in BPH and leaf folder prone areas/seasons
- Leave 30cm space at every 2.5m
- Use irrigation water judiciously
- Keep the fields free from weeds
- Use light traps to monitor pest incidence
- Remove egg masses of stem borer
- In brown planthopper (BPH) prone areas/seasons, avoid use of synthetic pyrethroids, methyl parathion and quinalphos and use only recommended chemical at recommended doses.
- Use insecticides based on economic threshold levels (ETLs).

Pests	ETL
Stem borer	2 egg masses/ m^2 or 10% dead hearts
Gall midge	10% silver shoots
Whorl maggot	25% damaged leaves
Green leafhopper	60/25 net sweeping or 5/hill at vegetative stage or 10/hil at flowering or 2/hill in tungro endemic area

Analysis of biodiversity and their natural enemies in cotton ecosystem and evaluation of IPM components

Cotton is an important fibre crop in India. The crop is damaged by more than 160 sp. of insects at different growth stages. Among these; aphids, leafhoppers, whiteflies and boll worms are important. The integration of bio control agents viz., parasitoids, predators and insect pathogens form an important component in the integrated pest management proform in cotton.

In cotton ecosystem, more than 150 species of natural enemies are found to exist. Besides, a few species of exotic natural enemies *viz.,* The egg parasitoid, *Trichogramma brasiliensis,* the larval parasitoid, *Bracon cirkpatriki* and the egg larval parasitoid, *Chelonus blackburni* attack cotton bollworms are established in several parts of India.

The following species of parasitoids and predators can be considered and promoted for use in the bio control of cotton pests.

3.2 Parasitoids

- *Trichogramma* spp: It is a cosmopolitan species and most popular hymenopterns egg parasitoid of large no of insects, especially lepidopterans. Adults are minute wasps, lays eggs on the eggs of *Helicoverpa armigera, Earias* sp, *Pectinophora gossypiella, Spodoptera litura* etc.

Rate of Release	:	1 lakh/ac per release during 8th, 9th,10th week after sowing
Time of Release	:	Early morning hours and avoiding release in rainy days
Method of Relese	:	Preparasitized Trichocards are stapled tied to the 4th leaf from top; eggs facing the earth; away from border rows.

- *Chelonus blackburni* : Belongs to egg larval parasitoid female adults parasitise the eggs and later develop inside the larval and emerge out killing the larva.

  The target insects under this group of parasitoid include *Helicoverpa armigera, Earias* spp. *Spodoptera litura* and *Pectinophora gossypiella*.

  Rate of release : @ of one adult/plant totalling 40,000/ha

  Time of release : Early morning and evening hours.

  Method of release : Adult parasitods are as such released by walking into the field by carrying the opened parasitoid containers.

- *Bracon cirkpatriki* : Parasitic on the larvae of *Earias* sp.
- *Telenomus remus* : Specific parasitoid on the eggs of *Spodoptera litura*.

3.3 Predators

- *Chrysoperla carnea* is commonly referred as green lace wings. The adults are free living and larvae are predatory on the eggs and early instar larvae of lepidopterans and all soft bodied insects. The target insects include aphids, thrips, whiteflies, eggs and larvae of all lepidopterous pests of cotton.

  Rate of Release : 40,000 larvae (Grub) (Preferably 2nd instar) per acre.

  Method of Release : Larvae along with paper bits are carried in polythene bags and are distributed inside the field in the early morning or evening hours avoiding the bright sunshine.

  During the release of parasitoids and predators in the cotton ecosystem, insecticide application shall be avoided at least 10 days prior and after release.

4. Schedule to be Followed for Cotton IPM with Special Reference to Biocontrol Agents

- Parasitoid: *Trichogramma* sp.
- Predator : *Chrysoperla carnea*
- Pathogens : NPV's and Bt

4.1 Steps

- Release of predatory grubs of *C. carnea* @ 40,000/ acre per release during 6th, 13th, 14th week after sowing.
- Release of egg parasitoid *T. chilonis* @ 1 lakh/ac per release during 8th, 9th and 10th week after sowing.
- Apply 2 rounds of *H. armigera* NPV 500 LE/ha at 7th and 12th week after sowing
- Apply HaNPV + 5% NSKE during 11th week after sowing SlNPV 250 LE/ha + crude sugar 2.5 kg to + 0.1% teepol could be administered.

4.2 Case Study in the Development of Biocontrol Based IPM in Cotton

Two experiments were conducted in a confounded block design, replicated 10 times with three sets of treatments in farmers field with MCU5 during winter season of 1992 and 1993 Each set of treatment was given in an area of 2000m2 which was subdivided into 10 subplot while set I was 150m away from set of treatment II and III.

The first set included the spraying of oxydemeton methyl 0.05% against sucking pests 30 DAS, release of *C. carnea* and 50,000 (2-3 day old larvae)/ha on 30, 37 and 44 DAS; Trichogramma chilonis @ 1,50,000/ha (8 times) at weekly interval from 51 DAS, need based application of pesticides Viz, endosulfan 0.07; monocrotophos 0.036% against bollworms, spraying of Ha NPV 500 LE/ha and neem oil 0.5% + teepol 0.1% or acephate 1.3kg/ha against whitefly.

The second set of treatments included 500 ml/ha oxydemeton, endosulfan 2l/ha, phasalon 2.5l/ha, quinolphos 2l/ha in alternation, Ha NPV 450 LE/ha and the third set included famous method using only insecticides. In general set I with bio control agents received only 5-6 sprays, set II received 6-9 sprays of farmers method included 10-14 sprays.

Observation of population of leaf hoppers, whiteflies, aphids and grubs of concinellids were released from 5 plants selected from random in each subplot. In each plant, sucking pests were recorded from 3 leaves from top, middle and bottom of plant at weekly intervals and mean population per plant was worked out yield data was recorded from each plant and converted per ha. The data were also worked out on cost-benefit ratio.

4.3 Results

- *Leafhoppers* : The population of leaf hoppers was re considerably in all the treatments. However, sets I and II were on par in recording less population on 51 DAS when compare to formers method.
- *Thrips* : Population gradually decreased from 23 to 48 DAS in all treatments and set I was superior for sets II and III on 58 DAS in recording lesser population.
- *Aphids* : Population was brought under check from 30 to 51 DAS by all sets of treatments set I and II were on par on 51 DAS when compared to set III.
- *Bollworms* : Sets I and II invariably recorded less Bollworms damage when compared to set III. The effects of treatments was well pronounced under set I (7.5%) following by set II (15%). When compared to 28.6% large in farmers method, inspite of increased number of application of insecticides.

5. Conclusion

Analysis of natural enemies and pest population in the rice field becomes necessary to convince the farmers to reduce the consumption of insecticides and also enable them to utilize the IPM practices.

References

Greathead, D.J. 1979. Critical review of natural enemies of insect pests on rice in South and South East Asia and their potential for biological control: Proposals for biological studies to assist in development and integrated pest control in South and South East Asia. Commonwealth Institute of Biological control, Slough, England.

Heong, K.L. 1991. Management of the brown planthopper in the Tropics pp.269-279. In: Migration and Dispersal of Agricultural insects, Tsukuba, Japan, September, 25-28, 1991.

Lim, G.S. 1970. Some aspects of the conservation of natural enemies of rice stemborer and the feasibility of harmonizing chemical and biological control of these pests in Malaysia. Mushi 43(11): 127-135.

Nickel, J.L. 1964. Biological control of rice stem borer: A feasibility study. International Rice Resarch Institute, Los Banos, Philippines, *Technical Bulletin*, 2.

Woodburn, 1990. Impact of pesticides in Farmer health and the rice environment: An overview of results from a multidisciplinary study in Philippines.

Yasumatsu, K. and T. Torii, 1968. Impact of parasites, predators and diseases on rice pests. *Annual Review of Entomology,* 13: 295-325.

❐❐❐

SECTION - III

UTILIZATION OF MICROBIALS

15

Utilisation of Microbials in Crop Protection

BAMBAWALE, O.M.
Director, National Centre for Integrated Pest Management, ICAR,
IARI Campus, New Delhi

Biological control practices for direct protection of plants from pathogens involve the deployment of "entomogenous" or "entomopathogenic" fungi, bacteria, viruses and antagonistic microorganisms at the infection court before or after infection takes place. Fungi which have a parasitic or pathogenic association with arthropods, particularly the true insects (Hexapoda), and mites (Acrania) are termed as "entomogenous" or "entomopathogenic" fungi. Although hundreds, possibly thousands, of microorganisms have been shown to interfere with the growth of plant pathogens in the laboratory, greenhouse, or field, and to provide some protection from the diseases caused by them, so far only few microorganisms have been registered and are commercially available for use. These microorganisms include fungi: *Gliocladium virens,* for control of seedling diseases of ornamental and bedding plants; *Trichoderma harzianum / T. viride,* for control of several soilborne plant pathogenic fungi and *Trichoderma harzianum / T. polysporum,* for control of wood decays. The other commercially available microorganisms are bacteria: *Agrobactrium radiobacter* for use against crown gall; *Pseudomonas fluorescens*, for use against *Rhizoctonia* and *Pythium* damping-off of cotton; and *Bacillus subtilis*, used as a seed treatment. Although the actual use of these biological control products is rather limited, it is expected that these and other such products will find wide acceptance and will fill a real need in the not too distant future.

1. Entomopathogenic Fungi

With renewed interest in entomological biological control and integrated pest management, more research has been directed to the role of naturally occurring entomogenous fungi in the regulation or suppression of population of insect pests. In addition, entomogenous fungi are receiving more attention as artificially introduced microbial insecticides and also are being identified as key natural mortality factors in the ecology of many important invertebrate pests. It should be pointed out, however, that entomogenous fungi hold a relatively modest position as compared to bacteria, as microbial control agents today. There are approximately 750 species of fungi found through out the five major taxa comprising the division Eumycota. Eumycota have been identified as being entomogenous or having confirmed or suspected entomogenous spp. The major taxa are the Mastigomuycotina (zygosporic fungi), the Ascomycotina (sac fungi), the Basidiomycotina (club fungi) and the Deuteromycotina, better known as Fungi Imperfecti. The potential entomogenous fungi are given in Table 1.

Table 1: Entomogenous Fungi being considered as Potential Microbial Insecticides

Fungus	Infective stage	Insect host	Habitat
Beauveria	Conidia	Beetles Caterpillars	Foliage
Metarhizium	Conidia	Froghoppers Beetles	Foliage
Nomuraea	Conidia	Caterpillars	Foliage
Verticillium	Conidia	Aphids Whitefly Scales	Glasshouse Foliage

1.1 *Beauveria*

During the last 15 years, special attention has been given to the use of *Beauveria bassiana* as a component of an integrated pest management strategy against several agriculturally important pests including whiteflies and the Lepidopteran pests including *Helicoverpa armigera*. Generally, two treatments made at 15-day intervals with 1.5 kg/ha concentrated product of *B. bassiana* (3. 0 x 10^9 conidia /g) is effective.

1.2 *Metarhizium*

The green muscardine, *Metarhizium anisopliae* var. *anisopliae* and var. *major* like the white muscardine fungi have been studied for 100 years. However, promising field results with this fungus have occurred only within the last 15 years. The rhinoceros beetle (*Oryctes rhinoceros*), a serious pest in coconut growing regions of the South Pacific, has been successfully controlled by *Metarhizium anisopliae* var. *major*. *M. anisopliae* was successfully used to control spittel bug on sugarcane and in pastures for nearly eight years. As much as 50,000 ha in a single province was aerially treated at 50 l/ha with 6 x 10^{11} to 1.2 x 10^{12} conidia/l. Although *M. anisopliae* occurs in natural epizootics, introducing the fungus early in the season appears to keep spittel bugs bellow damaging levels. *M. anisopliae* has also been effective in fields trials against mosquitoes. It has reduced pasture cockchafer populations when applied as spore bait. In China, *M. anisopliae* was pathogenic to *Colasposoma* when applied to the soil at a rate of 75 kg/ ha. From 50 to 60% mortality of pecan weevil also has been obtained in the U.S. with applications of *M. anisopliae*. At this time, there is considerable interest in developing *Metarhizium anisopliae* into a commercial product in our country for the management of white grubs in sugarcane and for the control of jassids and termites and the program is expanding to include control of other insects.

1.3 *Nomuraea*

Epizootics of *Nomuraea rileyi* on lepidopteran pests occur regularly in agro-ecosystems in India and other countries. Extensive research has been conducted on the survival of the fungus in soybean fields and a model has been proposed for predicting the level of infection in velvet bean caterpillar population in Florida. Early close planting of soybeans and cotton, that influences microclimatic field conditions, also enhanced the development of *N. rileyi* generally occur after caterpillars have caused damage and too late to prevent losses in soybean yield, the prophylactic use of *N. rileyi* has been tested using high conidial doses with encouraging results. A dust formulation of

N. rileyi with 12.4 x 10^{10} conidia /g applied early in the season at 1.1 x 10^{13}/ acre to soybeans caused a premature epizootic. These studies suggest that a prophylactic treatment 2 to 4 weeks before economic levels of injury are anticipated may provide control of defoliating caterpillars.

Table 2: Microbials listed for registration, commercialization and use in the country

S. No.	Pathogen	Disease	Crop
Antagonistic to Plant Pathogenic Fungi:			
1.	*Trichoderma* sp.	Root rot	Oilseeds
		Wilt	Pulses, Oilseeds, Cotton
2.	*Gliocladium* species	Root rot	Vegetable Crops
3.	*Pseudomonas fluorescence*	Root rot	Cardamom, Pepper, Ginger, Turmeric, Carrot, Betelvine, Tomato
4.	*Bacillus subtillis*	Root rot	Vegetable Crops
Entomopathogens: Fungi			
5	*Beauveria bassiana*	Berry borer Shoot borer	Coffee Sugarcane
		Bollworm	Cotton
6	*Metarhizium anisopliae*	Borers	Rice
		Pyrilla	Sugarcane
		White grub	Groundnut
7	*Nomuraea rileyi*	*Helicoverpa armigera*	Cotton, Pulses
		Spodoptera litura	Groundnut, Castor
8	*Verticillium lecanii*	Aphids	Oilseed Crops
Bacteria:			
9	*Bacillus thuringiensis var. Kurstaki*	Boll worms	Cotton
		Fruit borer	Brinjal, Tomato Okra
		Diamond Back Moth	Cabbage, Cauliflower Pulses
		Helicoverpa armigera	
10	*Bacillus sphaericus*	Houseflies Mosquito	-
Virus:			
11	NPV of *Helicoverpa armigera*	*Helicoverpa armigera*	Cotton, Pulses Vegetables Groundnut, Maize
12	NPV of *Spodoptera*	*Spodoptera litura*	Cotton, Castor Groundnut, Tobacco, Tomato
13	GV of *Achaea janata*	Semilooper	Castor
14	GV of *Chilo infuscatellus*	Shoot borer	Sugarcane

1.4 *Verticillium*

Verticillium (=*Cephalosporium*) *lecanii*, reported from scale insects for some time, has become important as a microbial insecticide for use in glasshouses. A spore preparation containing 41 x 10^7 active spores /g either undiluted or as a 10% concentration (diluted with talc or water) was applied to citrus leaves infested with *Coccus hesperidum* L. and in the greenhouse, 85 to 100% mortality was achieved within 3 weeks. Although *V. lecanii* killed all stages of the brown soft scale, it was unable to complete its development because of lack of nutrients. Hence the fungus, unable to survive and spread within the scale population, must be repeatedly introduced.

2. Fungal Antagonists

Trichoderma sp., mainly *T. barzianum T. viride* are the most common mycoparasitic fungi. The latter fungus has been shown to parasitize mycelia of *Rhizoctonia* and *Sclerotium*, to inhibit the growth of many other fungi, for example, *Pythium, Phytophthora, Fusarium,* and *Heterobasidion* (Fomes), and to reduce the diseases caused by most of these pathogens.

Mechanisms of Action

The mechanisms by which antagonistic microorganisms affect pathogen populations are not always clear, but they are generally attributed to one of four effects:

- Direct parasitism or lysis and death of the pathogen.
- Competition with the pathogen for food.
- Direct toxic effects on the pathogen by antibiotic substances released by the antagonist.
- Indirect toxic effects on the pathogen by volatile substances, such as ethylene, released by the metabolic activities of the antagonist.

Many of the antagonistic microorganisms are naturally present in crop soils and exert a certain degree of biological control over one or many plant pathogens regardless of human activities. Humans, however, have been attempting; increase the effectiveness of antagonists either by introducing new and larger populations of antagonists, for example, *Trichoderma harzianum* in fields where they are lacking, and /or by adding soil amendments that serve as nutrients for, or otherwise stimulate growth of, the antagonistic microorganisms and increase their inhibitory activity against the pathogen. Unfortunately, although both approaches are effective in the laboratory and in the greenhouse, neither has been particularly successful in the field. New microorganisms added to the soil of the field cannot compete with the existing microflora and cannot maintain themselves for very long. Also, soil amendments, so far, have not been selective enough to support and build up only the populations of the introduced or existing antagonists. Thus, their potential for eventual disease control is quite limited. There are several cases of successful biological control of plant pathogens when the antagonistic microorganism is used for direct protection of the plant from infection by the pathogen.

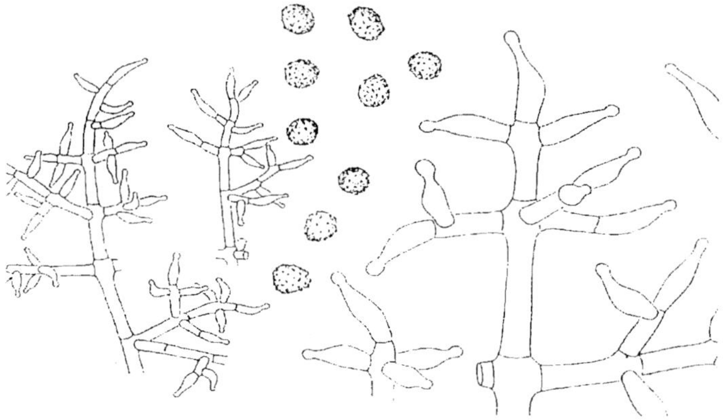

Trichoderma viride

2.1 Biological Control of Soil Borne Diseases

The principal fungi used as biological control agents against soilborne diseases are mainly, *Gliocladium virens* and *Trichoderma harzianum*. They are used in potting mixes, are mixed with soil, or are used and solid matrix in seed treatments. They are effective against damping-off diseases of ornamentals and vegetables caused by *Pythium, Phytophthora, Sclerotium,* and some other fungi. In addition, *Sporidesmium sclerotiorum, Conithyrium minitants, Talaromyces flavus,* and others give experimental control of some diseases caused by *Sclerotinia, Rhizoctonia,* and *Verticillium.* Some species of *Pythium*, such as *Pythium nunn* and *P. oligandrum,* protect potted ornamentals and vegetable from plant pathogenic species of *Pythium, and Talaromyces flavus* and binucleate Rhizoctonias protect plants from the pathogenic *Rhizoctonia solani.* Experimental control of several other soilborne diseases has been obtained with many other fungi. Finally, *Fusarium* wilts of several crops, such as celery, cucumber, and sweet potato, caused by the respective formae specialis for *Fusarium oxysporum* have been reported to be successfully controlled by inoculating transplants or cutting with nonpathogenic strains of the same fungus. Some of these strains have been isolated from the vascular tissue of host plants that remained healthy while nearby plants had been killed by the wilt-inducing strains of the fungus. It is believed that the nonpathogenic strains not only compete with the pathogenic ones in the rhizosphere and for infection sites, but they also enhance the resistance of the host toward the pathogenic strains.

2.2 Bacterial Antagonists, Pseudomonas Fluorescens

Pseudomonas spp. are ubiquitous Gamma subclass of Proteobacteria that are inhabitants of a wide range of soil, water and plant surfaces. Members of The bacterial genus Pseudomonas are Gram-negative, aerobic (able to consume oxygen) rods. Most are flagellated so they can move around. Most produce a slime layer that cannot be phagocytosed, and which aids in the production of surface-colonising biofilms. P. fluorescens is easily recognised as it secretes large amounts of a fluorescent, yellow-green siderophore under iron-limited conditions. Many Pseudomonadaceae are capable of producing these fluorescent pigments called pyoverdines. Growth of Pseudomonads is usually accompanied by a "fruity" odor compared to Escherichia coli. Certain strains of fluorescent pseudomonas are important biological components of agricultural soils that are suppressive to diseases caused by pathogenic fungi on crop plants. The biocontrol abilities of such strains depend essentially on aggressive root colonization, induction of systemic resistance in the plant, and the production of diffusible or volatile antifungal antibiotics. Evidence that these compounds are produced in situ is based on their chemical extraction from the rhizosphere and on the expression of antibiotic biosynthetic genes in the producer strains colonizing plant roots. Well-characterized antibiotics with biocontrol properties include phenazines, 2,4-diacetylphloroglucinol, pyoluteorin, pyrrolnitrin, lipopeptides, and hydrogen cyanide. In vitro, optimal production of

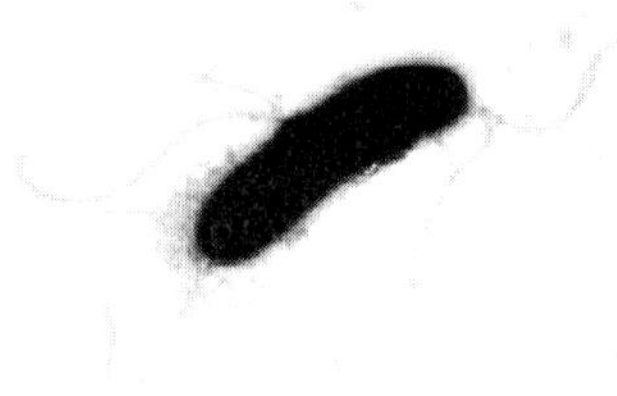

Pseudomonas fluorescens

these compounds occurs at high cell densities and during conditions of restricted growth.

Pseudomonas fluorescens encompasses a group of common, nonpathogenic saprophytes that colonize soil, water and plant surface environments. As its name implies, it produces a soluble, greenish fluorescent pigment, particularly under conditions of low iron availability. It is a obligate aerobe, except for some strains that can utilize NO3 as an electron acceptor in place of O2. It is motile by means of multiple polar flagella. P. fluorescens has simple nutritional requirements and grows well in mineral salts media supplemented with any of a large number of carbon sources. P. fluorescens is being researched as a biological control organism. Pseudomonas have an exceptional capacity to produce a wide variety of metabolites, including antibiotics that are toxic to plant pathogens.

2.3 Mode of Action

It is not clear exactly how the plant growth promoting properties of P. fluorescens are achieved; theories include:

- That the bacteria might induce systemic resistance in the host plant, so it can better resist attack by a true pathogen
- The bacteria might out compete with other (pathogenic) soil microbes, *e.g.* by siderophores giving a competitive advantage at scavenging for iron
- The bacteria might produce compounds antagonistic to other soil microbes, such as phenazine-type antibiotics or hydrogen cyanide.

2.4 Control of Diseases

P. fluorescens are well adapted to in soil, therefore, different strains are being investigated extensively for use in applications that require release and survival of bacteria in the soil. Chief among these are bioremediation of various organic compounds, and biocontrol of pathogens in agriculture.

P. fluorescens has been found effective in controlling fungal pathogens such as wilt/root rot Fusarium oxysporum f. sp. cubense, Pythium spp., R. solani, R. oryzae and S. rolfsii and bacterial pathogens like Xanthomonas citri and P. solanacearum in field tests. Bacterial preparations are widely being used in organic spice cultivation of southern India.

3. Bacteria

Bacillus thuringiensis (Bt) is a Gram Positive bacterium with unique insecticidal properties. It grows abundantly in the presence of oxygen (aerobic) and is a common soil inhabitant. Asiatic soils are particularly rich in Bt. In its life cycle Fig.1, Bt grows first as vegetative cells (rods of 2-5 ì). These rods are motile through the presence of appendages called flagella. One division of a rod into two rods takes approximately 20 minutes only. After abundant exponential vegetative growth in a favourable medium (organic matter) for about 12-24 hours, the rods enter into sporulation phase. A sporangium is produced through transformation of the vegetative rod which bulges in size and within this sporangium a spore on one side and a crystal (mostly rhomboid) on the other side is formed. The sporulation stage lasts for 36-72 hrs. At the end of sporulation the walls of sporangium disintegrate and the spores and crystals get separated out in the substrate. Upon receiving favourable conditions the spores germinate again to form vegetative rods and the cycle continues. The significance of crystals in the life cycle of Bt is not very clear. However, these crystals, called delta-endotoxin, which are protenaceous in nature, happen to be toxic to a wide range of insect pests. Bt commercial product essentially contains a mix of spores and crystals.

4. Nucleo Polyhedro Virus (NPV)

Nucleo Polyhedro Virus (NPV) is an important biological option of IPM available for H. armigera management because of its well demonstrated proven efficacy on a number of crops all over the country as also its several attractive features such as specificity, ecological safety, recyclability, less chances of resistance development by the pest and long shelf life. The option remains grossly underutilized for want of a proper understanding of its weaknesses and strengths as regards to the ecological niches where it can be utilized and the lack of organized effort for its production and use. The efforts made by several SAUs and Central Institutes in the nineties towards promotion of commercial production of Ha-NPV have not yielded the desired results, as the product is not yet available readily in the market. The option, therefore, could not prove itself.

The demand of HaNPV on just a few crops like cotton, maize, gram, pigeon pea, vegetables etc. runs into several crore larval equivalents (LEs). We need to establish several hundred such production units to meet the huge demand. The cottage enterprise has high potential in generating rural employment. There is an urgent need to take up the production of Ha-NPV on a large scale, otherwise, the option would remain only of academic interest.

❑❑❑

16

Registration for Mass Production of Microbial Biopesticides

VIMALA DEVI, P. S.
Principal Scientist (Agril. Entomology), Crop Protection Section,
Directorate of Oilseeds Research, Rajendranagar, Hyderabad- 500 030.

Plant protection against pathogens, pests and weeds has been progressively reoriented from a remedial approach to a rational use of pesticide chemicals in which consumer health and environmental preservation prevail over any other productive or economic considerations. Microbial pesticides have been introduced in this new scenario of crop protection and currently several beneficial microorganisms are the active ingredients of a new generation of microbial pesticides or the basis for many natural products of microbial origin. The development of a microbial pesticide requires several steps to be addressed right from its isolation in pure culture and screening by means of efficacy bioassays performed *in vitro, ex vivo, in vivo,* or in pilot trials under real conditions of application (field, greenhouse, post-harvest). For the commercial delivery of a microbial pesticide, the bio-control agent must be produced at an industrial scale (fermentation), preserved for storage and formulated by means of biocompatible additives to increase survival and to improve the application and stability of the final product.

Microbial insecticides are naturally occurring and genetically modified insect pathogens that directly or indirectly reduce insect pest populations. They include bacteria, viruses, protozoa, fungi, nematodes, and other organisms that infect and exterminate insects. The bacterium *Bacillus thuringiensis* is perhaps the best-known example. Because of the nature of biological pest control agents, some data requirements would have to be different from those necessary for registration of chemical pesticides, but the general principle that the product should demonstrate effectiveness and will not present unacceptable hazard to users, consumers of treated foods, or the environment still applies.

The main rationale for the development of microbial insecticides is environmental safety. Some pathogens selected for commercial development, such as viruses and bacteria, may infect only a single or small number of closely related insect pest species. Others, such as fungi and nematodes, may affect a fairly wide range of insects and related arthropod pests. However, none of these commercially available microbial pathogens have been shown to infect vertebrates or plants.

The data requirements for microbial pesticides are designed to give basic hazard and exposure information for a micro-organism with totally unknown properties. In actual practice, present microbial pest control agents are usually well identified which enables a regulatory authority to predict their properties and behaviour. This is particularly true in the categories of human health and plant pathogenicity. Clinical medicine and agricultural science have identified many micro-organisms associated with many diseases. If the microbial pesticide under consideration is taxonomically similar to a clinically or agriculturally-significant micro-organism, its properties and effects should be examined in greater detail than suggested by the tests generally required.

The import, manufacture, sale, transport, distribution and use of microbial pesticides are regulated in India under the Insecticides Act, 1968 and rules framed there under. Normally the Registration Committee (RC) constituted under the Insecticides Act, 1968, allows for commercialization in terms of import, manufacture, sale and use etc. of pesticides only after grant of regular registration under the Act. However, to ensure early availability of microbial pesticides to the farming community, the RC has simplified the registration procedure and allowed commercialization during the provisional registration of microbials.

The data requirements for microbial pesticides include:

1. Identity of the Product

To establish the identity and biological purity of microbial pesticides, the following detailed information on Chemistry must be supplied:

1.1 Active Agent

- Physical and chemical properties
- Systematic name and strain for bacteria, protozoa, fungi, etc. common name
- Natural occurrence of the organism, its relationships to other species, and history
- Manufacturing process
- Appropriate test procedures and criteria used for identification, such as morphology, biochemistry, and/or serology
- Composition of unintentional ingredients, their nature and identity, and content of extraneous organisms
- Methods of analysis

1.2 Finished Product

- Physical and chemical properties
- Quantity of active agent
- Name and type of formulation
- Nature and quantity of diluent
- Identity of non-active ingredients such as UV protectors, water retaining agents, etc.
- Stability of product and effect of temperature and storage conditions on biological activity
- Methods of analysis
- Shelf-life

2. Biological Properties of the Active Agent

- Information on which species are attacked, degree of specificity for target pest(s), genetic stability of the agent, areas and circumstances in which the agent is naturally occurring, and likely biological effects arising from use, should be supplied. The information should include:

- The natural occurrence
- The target host species of the pest and the pathogenicity or antagonism to that pest
- The infective dose level, transmissibility and mode of action
- Types of crops or premises to be protected
- Manner, rate and frequency of application

3. Toxicology Data

It should be established that the active agent is not a known pathogen of man or other mammals and that the preparation does not contain such pathogens as contaminants or mutants as determined by acceptable tests.

3.1 *Primary toxicology Data* (Single Exposure Studies)

- Oral toxicity/pathogenicity
- Dermal toxicity/pathogenicity
- Inhalation toxicity/pathogenicity
- Primary skin irritation
- Mucous membrane irritation
- Allergy/sensitization/immunosuppression
- Primary eye irritation

3.2 Eco-Toxicity

It is important to assess ecological risks due to microbial pesticides. In order to do this, their degree of species specificity and adverse effects on non-target species must be given careful consideration. The following data are required to establish basic potential effects:

- Toxicity to fish
- Toxicity to honey bees
- Toxicity to earthworm
- Toxicity to silkworm

4. Container Content Compatibility

5. Labeling and Packing

Microbial pesticides included in the schedule of Indian Insecticide Act 1968:

Bacillus species	*Bacillus* species (*B. sphaericus*, *B. thuringiensis*)
Bacillus subtilis	*Bacillus subtilis* cohn
Gliocladium Species	*Gliocladium* Spp
Pseudomonas Spp	*Pseudomonas* spp
Trichoderma Species	*Trichoderma* spp.
Beauveria bassiana	*Beauveria bassiana*
Metarrhizium anisopliae	*Metarrhizium anisopliae* var *acridum*
Nomuraea rileyi	*Nomuraea rileyi*
Verticillium lecanii	*Verticilium lecanii*
Granulosis Viruses (GV)	Granulosis Viruses (GV)
Nuclear Polyhedrosis Viruses (NPV)	*Helicoverpa armigera* NPV, *Spodoptera litura* NPV, *Neodiprion sertifer* NPV and *Neodiprion lecontei* NPV
Hirsutella species	
Verticillium chlamydosporium	
Streptomyces griseoviridis	
Streptomyces lydicus	
Ampelomyces quisqualis	
Candida oleophila	
Fusarium oxysporum (non pathogenic)	
Burkholderia cepacia	
Coniotyrium minitans	
Agrobacterium radiobacter strain 84	
Agrobacterium tumefaciens	
Pythium oligandrum	
Erwinia amylovora (Hairpin protein)	
Phlebia gigantean	
Paecilomyces lilacinus	

Penicillium islanidicum (for groundnut)

Alcaligenes spp.

Chaetomium globosum

Aspergillus niger-strain AN27

VAM-Vesicular arvescular mycorrhijae (Fungus)

Mycothecirem verrucaria

- *Photorhabdus luminescences akhurustii* Strain K-I

Serratia marcescens GPS 5 (Bacterium)

❑❑❑

17

Impact of Research and Development (R & D) in Enhancing the Field Potential of Baculoviruses

NARAYANAN, K. AND G.P. SHETTY
Multiplex Bio-Tech Pvt. Ltd., Bangalore – 560 086, Karnataka

1. Physico-Chemical Approaches in the Use of Baculoviruses

1.1 Cultural Approaches: Timing of Application

Since viruses are rapidly inactivated by sunlight, it is often been suggested that they must be applied at dusk rather than earlier in the day to delay this rapid loss in activity. In general in most of the noctuid insects, not only the quantum of feeding but also the rate of feeding is more in night hours compared to day time. Further it was reported by Narayanan and Jayaraj (1988) that acquisition-feeding period of 12-18 hrs, by *Helicoverpa armigera* caused more than 80% mortality of *H.armigera* when HaNPV was applied during evening hours after 4 pm.This was substantiated from the results of field trial, wherein there was significant difference in yield of Bengal gram seed, when HaNPV virus was applied @ 250 LE/ha during evening hours when compared to morning application (Narayanan, 1980).

1.2 Directional Spray

Virus application may vary depending upon the feeding habits of the pest and the crop, whether it is on to the upper surface or under surface of leaves. For e.g. virus deposited on the upper surface of leaves is not available to them until late instars, and by then much of the virus will have been inactivated by sunlight. Therefore applications must be directed against most susceptible stages of the insect pest i.e. against early larval stages so that death occurs before the larvae reach more damaging developmental stage; targeting under surface of leaves thereby increasing the efficacy of baculoviruses. Timing of application for a virus-host complex may also vary with plant host. For e.g. larvae of *Heliothis* spp. on cotton feed on concealed pods much of the time, and application of virus must be directed against very young larvae that feed on new terminal growth before moving down into fruiting structures. On the other hand, such as grain sorghum, and sunflower, larvae feed on exposed on the open fruiting heads and timing is less critical (Jayaraj *et al.,* 1985). (Table)

1.3 Spray Fluid Volume

Although various viral formulations have often been applied either aerially in forest or ground application in Agricultural row crop situations, very few studies have been conducted in the requirement of spray fluid volume in relation to efficacy. It has been stated by Chapman and Ignoffo, (1972) that doubling the volume of spray fluid from 94 to 188 l/ha provided control equal to doubling the rate of *H.zea* NPV from 102 to 204 LE/ha in the case of high volume sprayer like knap-sack sprayer.

1.4 Formulation with Different Additives

Among several factors which limit the efficacy of baculoviruses in the field, the rapid inactivation of viral occlusion bodies by solar radiation is the most important one. Therefore formulation can play an important role in stabilising the pathogens. Formulation with different additives can substantially improve the efficiency of the application, persistence activity of the virus in the field. The need for improved performance specifications of viral formulations has led to the development of a variety of adjuvants for addition at application to tank mixtures containing viral sprays. Additives have been developed for a variety of purposes, i.e., increasing stability of viruses on crops, gustatory stimulation of larval hosts, improved droplet characteristics and coverage and evaporation retardation. Further they are selected based on the significant improvement in the desired physical properties of baculovirus. A lot of study has been conducted with different additives both to the basic and tank mixture formulations of baculoviruses for increasing their persistence and efficacy (Narayanan, 2002). For want of space the results of those studies conducted in India alone are summarised a presented in tabular form (Table 1).

Table 1 : Additives that have been added to the formulations to improve the bioefficacy of baculoviruses in India.

Kind of additives	Name of the Virus/crop Tested	Remarks
Diluent (Dust) Talc, Kaolin, Clay	HaNPV	L/M/S
Wetting agent/ Spreader Triton X-100	HaNPV/Chickpea	F/Y/S
Dedinol /Lissopal D/Lissopal GM.	SlNPV	P/M /NS
Adhesive and sticker		
Whole milk	HaNPV, Cotton	F/Y/S
Sugarcane jaggery	HaNPV, Sunflower	F/Y/S
Baits		
Cotton seed kernel/Chickpea/ tapioca flour	HaNPV	P/M/S
Groundnut/Cottonseed/Palm oil	HaNPV,	P/M/NS
Miscellaneous		
Sucrose/Lactose, Crude sugar	HaNPV,	F/Y/S
Larval extract	HaNPV	L/M/S
Tender coconut water	HaNPV	L/M/S
Egg yellow and egg white	HaNPV	P/M/S
Sun light protectant		
Robin blue /Tinopal	HaNPV	P/M/NS
Congo red	HaNPV, SlNPV	L/M/NS
Charcoal /Black ink	SlNPV, Cauliflower	F/M/NS
Biopolymers		
Alginate /Starch	Ha NPV/ Sunflower/ Redgram	F/Y/S
Multiple purpose additive		
Shade (phagostimulant, evapo-retardant and sunlight protectant	HaNPV, Chickpea	F/Y/S

SlNPV= *Spodoptera litura* NPV HaNPV=*Helicoverpa armigera* NPV
F=Field study L= Lab study P=Pot/cage study M= Mortality Ns=Non significant
S=Significantdifference either in larval mortality/yield Y=Yield

1.5 Compatibility with other Pesticides

Crops often require an array of pesticides during a growing season for control of diseases, weeds, insects etc. For control of diseases, weed, insects, etc. it is often to treat two or more of these pest problems simultaneously by applying a mixture of pesticide in a single application. It is thus advantageous for viral insecticide to be compatible in tank mixture with as wide a range of pesticides as possible. Viral insecticides have been found to be compatible in tank mixture with most conventional synthetic organic insecticides like is organochlorine, organophosphorus, carbamate, and pyrethyroids insecticides. Other pesticides with which *H.zea* NPV has been found compatible in sprays are chlordimeform and cyhexatin (Acaricides), Thiabenzdole and Benomyl (fungicides), and methoprene, diflubenzuron (insect growth regulators). *Plodia interpunctella* GV was compatible with the fumigants phosphine, carbon tetra chloride, carbon bisulphide or ethylene dichloride, but not with methylbromide.

1.6 Use of Optical Brightener

Two major drawbacks that have hampered the effective use of baculoviruses for pest management are (i.) loss of infectivity in the environment by way of being susceptible to ultraviolet (UV) radiation and (ii) protracted time to death *i.e.,* slowness in causing lethal. Recent studies described the use of optical brightener as a radiation protectant for the gypsy moth *Lymantria dispar* MNPV as well as an enhancer by way of enhancing the residual activity upto 214 fold.

2. Biological Approach

2.1 Strain Selection by Geographical Isolation

(1) Phenotypic variation among different isolates also has been reported for many NPV's in terms of survival time, virulence and pathogenicity. Pathogenicity to NPV's with reference to their migration is also available.Viruses from members of the same host species from different geographical locations show variability in virulence and biological characteristics and their DNA differs in restriction endonuclease and SDS-PAGE protein patterns. Variants of baculoviruses with heritable variations in virulence and host range arise spontaneously in nature. Field isolates collected in diverse geographic locations are often genetically heterogenous.Differences in biological activity to the tune of 16 fold has been detected among

different isolates of gypsy moth, *Lymantria dispar* collected from North America, Europe and Asia.

2.2 Selection for UV Resistance

A strain of *Cydia pomonella* GV which was 5.6 times more resistant to UV light than the original isolate and remained infective for twice as long in the field was selected. Such investigations in our agroecosystem are bound to yield fruitful results, as there is a greater biodiversity under tropical conditions.

2.3 Following Serial Passage

Selection of variants with increased virulence is an attractive option for increasing the utility of baculoviruses. When, wild type *Autographa californica* multiple embedded nucleo polyhedrovirus (Ac MNPV) was serially passaged 20 times in third instar Diamond backmoth of cabbage, *Plutellla xylostella*, the virulence of passaged AcMNPV was increased approximately 15- fold in contrast to parental wild isolate of AcMNPV. An isolate of *Choristoneura fumiferana* NPV was fed to neonate larvae of the cabbage looper and the wax moth, after passage, the virus was able to infect new hosts.

2.4 *In vivo* Production

At present several baculoviruses are produced commercially by the insectory method, using artificial diets with little or no sterility precautions. The recent advances in laboratory culturing of insect cells might offer a satisfactory alternative method for large scale production of baculoviruses.Obligate pathogens like baculoviruse require a living insect host for mass production. In some case pathogen can be mass produced in a factious host.

2.4.1 Preferential Use of Females

The objective of virus production is to utilize host tissue as efficiently as possible and to obtain the greatest amount of biologically active virus possible. In general, late stage larvae have been utilized for virus production, since more viruses are produced in these insects than in young insects. In general female larvae in lepidoptera invariably have one more instar than the male and weighs considerably. Hence, possibility of preferential use of female larvae for large scale in vivo production of baculovirus using lepidopteran bio-mass effectively has been attempted.

2.4.2 Alternate Hosts

In general, entomopathogenic viruses have been produced in their respective homologous hosts, presumably because of their host specificity. Many of these insects, however, may have undesirable qualities for rearing and virus production, ie, obligatory diapause, prolonged developmental time, presence of allergenic and urticarious setae, small size, cannibalistic behaviour, etc so that more 'desirable' hosts might be sought.Virus produced in the alternate or factitious host, but there may be great difficulties in either approach. Most of the work on virus specificity or host range concerned the insecticidal activity of the virus. Several noctuid species were exposed to a multicaspid NPV of *Heliothis armigera.* While neonate larvae of the bollworm were most susceptible (LC_{50} = 10^3 POB per milliliter of diet), larvae of the tobacco budworm and the beet armyworm (*Spodoptera exigua*) were less susceptible (LC_{50} = about 10^5 - 10^6 PIB per milliliter. Since *H. zea* NPVwas found to be infective for neonate larvae of *H. armigera* and *H. punctigera,* it was felt that *H. zea* NPV could be considered for control of the other *Heliothis* species in Australia, because of its demonstrated safety and production technology. Very recently Narayanan (2003) has reported the cross infectivity of granulovirus of field bean pod borer, *Adisura atkinsoni* to *H. armigera.* This can be advantageously utilized for the large-scale multiplication of GV in *H. armigera* in the laboratory in view of its short larval period of 12-14 days and large biomass (407 mg) without undergoing any dormancy, when compared to *A.atkinsoni* which has a very long larval period of 20-22 days with average weight of 350 mg and undergoing pupal diapause.

2.4.3 In vitro Production of Baculovirus

Tissue or cell culture techniques for the mass production of obligate pathogens like insect baculoviruses have not yet materialized. Although viruses produced in this manner are effective under field condition. The cost of utilizing cell culture for mass producing insect viruses at this time is prohibitive. However, advances are being made in this field and perhaps some day this technology will be available for the mass production of insect baculoviruses.

2.4.4 Integrated with other Pathogens: Entomopox Virus

Most entomopoxviruses (EPVs) form two distinct types of proteinaceous structure viz; spindle and spheroid. The spheroid contains virions. On the other hand, the spindle composed mainly of the protein fusolin, has no virions. Recent evidence indicates that the spindles associated with the spheroids of *Pseudaletia seperata* entomopox virus can enhance the effectiveness of a nuclear polyhedrosis virus by way of reducing the LD_{50} of the NPV of *P. seperata,* infecting the same host by as much as 9000 times. Very recently, have shown that the enhancement of non-occluded viruses of *B. mori* nucleopolyhedrovirus and polyhedrin negative recombinant (BmNPV) when the silkworm larvae are administered orally with *Anomala cuprea* entomopoxvirus spindles. Further the future possibility of enhancing the nucleopolyhedrovrus of oriental armyworm, *Mythimna separata* using spindles from *H. armigera* entomopox virus with an enhancement index ($\log_{10}$) of the virus was found to be 2.76 or reduction of more than 500 times in Lc_{50} has been reported.

2.5 Integrated with Granulovirus

Enhanced infection of a nucleopolyhedrovirus in larvae of the oriental armyworm by the capsules of granulovirus isolated from *Spodoptera litura* was reported by Chakraborty *et al* (2005). The enhancement index ($\log_{10}$) of the virus was found to be 3.33 or reduction of nearly 2000 times in LC_{50} value.

2.5.1 Integrated with other Nucleoployhedrovirus

Recently have shown the enhanced infection of a nucleopolyhedrovirus in larvae of the oriental armyworm, *M separata* by the spindles shaped inclusion bodies isolated from an heterologous NPV of *Galleria mellonella* NPV. The enhancement index ($\log_{10}$) of the virus was found to be 3.33 or reduction of more than 3450 times in LC_{50} value.

2.5.2 B.t.: Integrated with Bacillus thuringiensis (B.t.)

Granulosis virus infection in *H. armigera* increased the susceptibility of *H. armigera* larvae to *Bacillus thuringiensis* was lower in GV- inoculated larvae compared to the control.

3. Conventional Genetic Approach

3.1 Through Recombination by Mixing

In vivo recombination has been carried by mixedly infecting *Galleria mellonella* larvae with two closely related baculoviruses *Autographa californica* MNPV and *Galleria mellonella* MNPV.

3.2 Through Chemical Mutagens

Mutants can be generated in tissue culture by growing the wild type virus in the presence of the mutagen. The spruce budworm virus was grown in the presence of the mutagen nitrosomethyl guanidine and the surviving virus was cloned. Of the 34 plaque isolated and examined, one isolate, CfNTG 29, was more virulent than the standard. A direct application of genetics to viral insecticides is to generate stable variants of baculoviruses which are more effective in the field. Wood *et al.* (1981) could isolate a mutant strain of AcMNPV designated HOB which produced a large number of occlusion bodies in infected cells and which had higher virulence in insecte than the parent strain.

4. Biotechnological Approach

Even though several wild type baculoviruses have been investigated and utilized in biopesticide for row crop use for more than three decades, their wide spread use and acceptance has never been achieved due to the very slow kill which is characteristeic of wild type viruses. The advent of biotechnology provided a new opportunity to make these organisms efficient biopesticides by using them to deliver a foreign gene or its own gene or deleting some gene which would have a lethal impact or otherwise in increasing the efficiency and utilization of insect baculoviruses. Very late in the infection of baculovirus, two proteins, viz., polyhedrin and protein (P10) are abundantly synthesized. The only role of polyhedrin appears to be polyhedral occlusion body (POB's) formation, which is irrelevant or not essential for the production or multiplication of invitro infective non-occluded virus. Foreign gene can, therefore, be inserted in place of the polyhedrin gene eliminating its function. The resulting viral genome will initially direct the production of non-occluded virus and with the time course of polyhedrin gene expressed, go on to produce product(s) as directed by the inserted foreign DNA.

4.1 Insertion of Insect Specific Neurotoxic Foreign Genes: Middle Asian Scorpion Toxin Gene from – *Buthus eupeus*

Possibility of using recombinant DNA technology (r-DNA) to incorporate a synthetic gene encoding an insect specific paralytic neurotoxin found in the venom of Middle Asian scorpion *Buthus eupeus* into *Autographa californica* multiple embedded nuclear polyhedrosis virus (AcMNPV) genome with the aim of halting insect feeding more rapidly has been explored.

4.2 North African (Algerian) Scorpion Toxin Gene (AacT) from *Androctonus Australis*

A recombinant baculovirus derived from AcMNPV containing the venom of North Afrcian (Algerian) scorpion *Androctonus australis* to improve the virulence of AcMNPV by reducing the survival time of the infected insects with concomitant reduction in host plant damage has also been attempted.

4.3 Predatory Straw Itch Mite Toxin Gene from *Pyemotes tritici*

The improved insecticidal activity of the recombinant virus containing predatory straw itch mite toxin from *Pyemotes tritici* demonstrated by the reduction in both median survival time (ST_{50}) and median lethal dose (LD_{50}) value in *Trichoplusia ni* larvae when compared to wild AcMNPV (ST_{50}, 113 hr) which registered a decrease of 25 %. Increased paralytic effect of the novel neurotoxin encoded by tox34 whose molecular weight is 27 KD venom proteins, (commonly known as *txp-1)* isolated from the predatory mite *Pyemotes tritici* has also been reported. Further, they have improved *Heliothis zea* NPV by way of inserting mite toxin gene and expressed the toxin during infection resulting in 50 % mortality within 40 h after virus treatment. This is said to be the fastest acting recombinant baculovirus reported to date.

4.4 Parasitic Venom Gene from *Bracon hebetor*

Study by Quistad *et al.* (1994) on the purification and characterization of insecticidal toxin from the venom of the parasitic wasp, *Bracon hebetor* have shown the future possibility of using this toxin (*Brh*) as insecticidal through baculovirus vector system since it has shown 400 fold better higher biocidal activity (0.03mg/g) against *Spodoptera* when compared to scorpion toxin LD_{50} (13 mg/g) and 100 fold higher active against *Galleria* when compared to mite toxin (LD_{50} 0.3 mg/g).

4.5 *Bacillus thuringiensis* δ Endotoxin Gene

Delta endotoxin gene from *Bacillus thuringiensis* subsp. *kurstaki* HD 173 has been inserted into AcMNPV using both polyhedron and *p10* promoters. Analysis of infected cell extracts showed that the delta endotoxin was expressed in insect cells as 130k, 62k, 42k proteins with peak synthesis at 18 hr post infection. When extracts from the cells infected with polyhedrin negative virus were fed to *T. ni* larvae, feeding by the insect was inhibited and death occurred. Recently Chang, *et al.* (2003) have generated a recombinant *A. californica* NPV that produces occlusion bodies incorporating Bt Cry1Ac toxic protein where it is capable of causing the host mortality from one day post infection.

4.6 Maize Mitochondrial Protein URF-13

A maize mitochondrial protein involved in cytoplasmic male sterility and disease susceptibility has been expressed in AcMNPV. This protein, URF13, is hydrophilic and binds tightly to the membrane of cells. Injection of larvae of the cabbage looper *T. ni.* with polyhedrin-negative recombinant virus resulted in a 40% decrease in lethal time compared with that of the wild-virus.

4.7 Insertion of Activated Gene

4.7.1 Polydna Virus

Certain female parasitic wasps in the families of Ichneumonidae and Braconidae carry the particles containing double stranded circular, multiple DNA virus called Ichnovirus and Bracovirus. These viruses are injected through the cuticle during the oviposition of parasitoids. These are essential for the survival of their progeny in their habitual host. In the absence of these virus particles, the parasitoid egg is recognised as foreign and encapsulated by host blood cells *i.e.*, insect haemocytes, especially by granulocytes, whereas in the presence of virus, the parasitoid is not encapsulated. The incorporation of polydna virus genes with immuno suppressant activity into the genome of HaNPV in future might lower resistance in grown up larvae of *H. armigera* and enable this pest to be controlled even in the grown up stage with a recombinant HaNPV as it has been shown recently by way of allowing *C. sinorensis,* an ichneumonid parasite to oviposit into *H. zea* larva, which is highly resistant to AcMNPV; immediately before the *H. zea* larvae were orally inoculated with AcMNPV.

4.7.2 Entomopox Virus

Hukuhara *et al.*, (1999) have introduced an entomopox virus gene encoding a virus, enhancing factor (EF) into rice, which resulted in high level accumulation of the EF in the transgenic rice plants. Their bioassay studies have shown higher susceptibility of *P. separata* larvae to its baculovirus when it was fed with transgenic rice.

4.7.3 Insertion of Insect Genes: Diuretic Hormone Gene

Water balance in insect is highly controlled by its intake and excretion in response to changes in the insect environment. Diuretic and anti-diuretic hormones are considered to play important roles in maintaining the water balance in insects. This balance might be disrupted if a recombinant virus produced elevated levels of either of the hormone in infected larvae. A synthetic diuretic hormone gene of the tobacco hornworm was expressed in a recombinant baculovirus viz., *B.mori* NPV. When silkworm larvae were injected with the resulting BmNPV, the diuretic hormone was expressed causing a strong alteration in the larval fluid metabolism. The haemolymph volume and mortality of *B.mori* larvae infected with recombinant BmNPV has been decreased by 30% in comparison with wild type NPV infected larvae, which required five to six days to achieve the same level of mortality.

4.7.4 Juvenile Hormone Esterase (JHE)

The regulatory process associated with the metamorphosis of insect larvae into pupae is also suitable target for the action of recombinant baculoviruses. The metamorphic changes of insect caterpillars into pupae, ultimately into adults are regulated by the juvenile hormone (JH). A reduction in the titre of JH early in the late instar has been shown to initiate metamorphosis and lead to cessation of feeding behaviour. The reduction in JH is associated with a drastic increase in the levels of juvenile hormone esterase (JHE). Considering the above concept, *jhe* gene has been cloned into the transfer vector of AcMNPV and produced the recombinant NPV carrying the gene and they have reported stunted growth and reduction in the final weight, by about five times in the cabbage looper *Trichopusia .ni* larvae infected with recombinant NPV compared to control.

4.8 Insertion of Viral Genes: Use of Baculovirus Genes

4.8.1 Enhancin Gene

The baculovirus genome contains a number of genes with varied functions. One of the viral genes that affect the pesticidal activity of the baculovirus is viral enhancing factor (VEF) or 'enhancin' gene of *Trichoplusia ni* granulovirus (GV).. Recently an invertebrate intestinal mucin (IIM) from a lepidopteran insect *T. ni*, has been identified which is similar to the mucus layer found in mammals, protecting the digestive tract from microbial infections. Further, It has been shown that *T.ni* granulovirus (TnGV) has evolved a novel strategy with the possession of a granulovirus (GV) encoded protein 'Enhancin' to overcome intestinal mucinous defense barrier against micro-organism by way of its mucin degrading activity thereby increasing the susceptibility of larvae to nucleopolyhedrovirus (NPV) infection and decreasing the larval survival time.

4.8.2 Cysteine Protease (V. cath gene)

A functionally active cysteine protease gene encoded by *B. mori* NPV and AcMNPV has been reported. Further it has been shown the role of cysteine protease (*v-cath*) gene in the breakdown of infected host tissue thereby facilitating the horizontal transmission of virus. Infection of insects with recombinant viruses containing an inactivated *v-cath* gene results in insect with an altered appearance (after death) that fail to disintegrate normally. Thus *v-cath* appears to facilitate the release of occlusion from insect after death.

4.8.3 Chitinase Gene

Unlike vertebrates, which have extensive exposure of epithelial cells to the external environment, insects have extensive protection of their epithelial tissue. The chitinous cuticle of the insect covers virtually all external surface, even extending through the foregut, hindgut and tracheal tubes, constituting the first line of passive defense in insects. Recently a functional chitinase gene (*chiA*) has been identified in the genome of AcMNPV and high-level expression of its endo and exo-chitinase activity under insect cell system has been reported. The median time for mortality of fourth instar larvae of *S. frugiperda* infected with a recombinant virus containing chitinase gene was approximately 20 h shorter than that for insects infected with a wild-type virus.

4.8.4 25k FP Gene

Just lie virus encoded chitinase and protease involvement in the liquefaction of virus infected insects. The 25K gene from *B.mori* NPV as an additional gene involved in the post-mortem host degradation has been identified. It alters cell metabolism and /or host cell gene expression, in addition to its role in lowering of polyhedron production, i.e. few polyhedra phenotypes (FP). Thus 25K gene was invoved in the complex phenotype induction by virus infection and was important for virus transmission. Without 25K gene, host degradation followed by efficient dispersal of polyhedra would not occur.

4.8.5 P74 Gene

A few gene products have been implicated in the insecticidal activity of baculoviruses or in their infectivity or pathogenicity. The p74 gene, located downstream from the p10 gene, is involved in establishing lethal fusion infection in columnar gut cells, despite the absence of prominent viral spike glycoprotein (gp64) which is usually present in budded virus. AcMNPV encoded p74 is essential for establishing lethal infection by oral inoculation with occlusion derived virus (ODV) but not by intrahaemocilic inoculation with budded virus (BV) has been domanstrated. Deletion of this gene abolishes infection of insects whereas the overproduction of p74 enhances infectivity five-fold.

4.8.6 P 143 Gene (Putative DNA Helicase Gene)

Although baculoviruses infect over 600 species of insects, individual isolates normally show a limited host range and infect only closely related species. Thus, baculoviruses in general have narrow host specificity. This property makes baculoviruses potentially very useful biocontrol agents for the control of insect pests. To increase their usefulness as effective biological insecticides, it is desirable to alter these viruses genetically to be more virulent and to infect a broader host range. The putative DNA helicase gene *p143* has been shown as essential for DNA replication. Ahrens and Rohrmann (1996) reported the sequence of helicase gene. *Bombyx mori* NPV (BmNPV) and AcMNPV infect *Bombyx mori* and *S. frugiperda* cells respectively, but cannot replicate in the heterologous cell line. However, in co-transfection experiments AcMNPV recombinants containing a fragment of the BmNPV helicase gene instead of the corresponding AcMNPV sequence were produced that were capable of replicating in both *B. mori* and *S. frugiperda* cells.

4.8.7 Apoptotic Suppressor Gene (p35 & Iap)

ACMNPV carried apoptosis gene *p35* in addition to *'iap'* gene and both the genes together inhibit apoptosis in a wider range of insects. This could explain the relatively broad host range exhibited by AcMNPV compared to other baculoviruses with a narrower host range. In the case of *Orgyia pseudotsugata* NPV since it lacks *p35* gene, it has a narrower host range though it is closely related to AcMNPV.

4.8.8 Host Range Factor Gene (hrf-1)

The broad spectrum AcMNPV does not infect either gypsymoth larvae or its cell line IPLB-ld 6527. Recently Thiem *et al,* (1996) have identified a host range factor (*hrf-1*) gene that promotes AcMNPV replication in IPLB-ld 6527 cell. Co-trasfection AcMNPV and plasmid carrying the ldMNPV gene into IPLB-ld 6527 cells results in AcMNPV replication.

4.8.9 Host Cell Specific Factor Gene (hcf-1)

A gene which has got differential effect on late gene expression in two different cell line viz. Tn-368 and Sf-21 hence they named it Host Cell Factor (*hcf-1*) has been identified. Later the same authors characterized the effects of deleting the AcMNPV hcf-1 genes (*hcf-1* mutant) on the virus infection of different *T.ni* cell lines and *S. frugiperda* and *T.ni* larvae. It demonstrates that hcf-1 functions in a species and tissue specific manner and is important for AcNPV virulence in *T. ni* larvae. The above in vitro and in vivo studies with hcf-1 mutants shows that hcf-1 is a species specific virulence factor of AcMNPV. The existence of such species specific factors in baculovirus genomes implies that the ability of the viruses to infect alternate species may be an important aspect of their natural life history. Manipulation of such genes may allow the control of their distribution in nature as suggested by the authors.

4.9 Deletion of Gene

4.9.1 Ecdysteroid UDP – Glucosyltransferase Gene (egt)

Ecdysteroid UDP-glucosyltransferase (*EGT*) produced by AcMNPV maintains the insects in an actively feeding state throughout the infections by blocking moulting. The deletion of the gene encoding EGT was thought to accelerate the virus induced mortality by allowing the infected larvae to begin moulting, resulting in feeding cessation. In contrast with all other earlier strategies, all of which involved the expression of some foreign genes; the '*egt*' gene has been deleted, which significantly improves the pesticide characteristic of AcMNPV. Larvae infected with '*egt*' deletion mutant display considerable reduced feeding (287 mg) as compared to uninfected (402 mg) and earlier mortality (ST_{50} 99.7) as compared to (ST_{50}127.2) wild with LC_{50} of 2.39 and 1.42 h respectively.

4.9.2 Polyhedral Envelope – Associated Protein (pp. 34)

Polyhedral envelope associated protein (pp.34) which normally enveope the polyhedral occlussion bodies. Deletion of the gene responsible for pp. 34, makes the virions easier to find their way to midgut epithelium. So higher initial hit rate and accordingly less requirement of POBs for the establishment of infection.

It is obvious from the above discussion that various approaches, like, physico-chemical, biological including biotechnological approaches led to the development more effective, faster acting and better utilization of insect viruses. Further, it is evident from the above examples, that genetic manipulation of insect viruses apart from the conventional genetic approaches, can be modelled for specific and broad range of insect control with an aim to form "Super organism" with high virulence, pathogenicity, host range, persistence and mass production. Though some success of using indigenous baculovirus like *Heliothis*, *Spodoptera*, *Chilo* spp and *Oryctes rhinoceros* through conventional approaches has been achieved in India in order to compensate for the high costs of chemical insecticides; development of selective recombinant insect viruses is a must which will augment any further integrated pest management programme by reducing the impact on non-target species. Out of 1,100 host record covering 13 insect orders belonging to 12 viral families world wide including nearly 488 isolates of nucleopolyhedro iruses India has more than 50 host records covering 3 insect orders belonging to 5 viral families.

Majority of work that has been carried out so far relates to the baculovirus of *A. californica* both as biopesticides as well as baculovirus expression vector system. Though in India some preliminary work has been done towards molecular characterization of certain indigenous baculoviruses not even single studies has been conducted on the genomic analysis of any lepidopteran insect viruses in India when compared to more than ten insect viruses whose genomic analysis has been thoroughly studied elsewhere. Hence, the need of the hour is "Back to the Basic" approach to initiate more studies on the various aspects of molecular studies on different indigenous insect viruses including using of recombinant DNA technology for effective utilization various insect viruses for the microbial control of insect pests in India.

References

Chakraborty, M., K. Narayanan, and M.K. Sivaprakash, 2005. Enhancement of *Mythimna separata* nucleopolyhedrovirus with viral enhancing factor from *Spodoptera litura* granulovirus. *Biopestic.Int* I:109-113.

Chapman, A.J., and C.M. Ignoffo, 1972. Influence of rate and spray volume of a nucleo polyhedrosis virus on control of *Heliothis* in cotton.. *J. Invertebr. Pathol.* 20 : 183-186.

Hukuhara, T., T. Hayakawa and A. Wijonaiko, 1999 Increased susceptibility of armyworm larvae feeding on transgenic rice plants expressing an entomopox virus gene. *Nature Biotechnol.* 17: 1122-1124.

Jayaraj, S., R.J. Rabindra and K. Narayanan, 1985. Development and use of microbial agents for the control of *Heliothis* spp. (Lep: Noctuidae) in India. pp. 483-504 in *Proceedings of the workshop on biological control of Heliothis: increasing the effectiveness of natural enemies.* Ed. E.G. King and R. D. Jackson. Rekha Printers Pvt. Ltd.

Narayanan, K. 1980. Field evaluation of nuclear polyhedrosis virus of *Heliothis armigera* (Hbn.) on chickpea, *Cicer arietinum* L. Biological Control of *Heliothis* Workshop, 23-24 September 1980. Department of Primary Industries, Tom Street, Toowomba, Australia.

Narayanan, K. 2002. Current Status of Baculovirus Formulations and Application in India. ICAR-CABI Workshop on Biopesticide formulation and Application. PDBC, Bangalore, Dec 9th –13th , 2002.

Narayanan, K and S. Jayaraj, 1988. Effect of acquisition feeding period on the mortality of *Heliothis armigera* (Hbn.) due to nuclear polyhedrosis virus. *J. of Biol. Cont.*, 2: 135-136.

❑❑❑

18

Mycoinsecticides : Towards Organic Pest Control

NARAYANASAMY, P.
Dean, Faculty of Agriculture & Professor & Head, Department of Entomology, Faculty of Agriculture, Annamalai University, Annamalai Nagar-608002, Tamil Nadu

1. Introduction

The interest in microbial pesticides is largely due to emerging many problems associated with the extensive use of chemical pesticides. Not only do chemical pesticides generally affect beneficial insects as well as pest species, but insects tend to acquire resistance to the chemicals so that new pest problems rapidly develop. Furthermore, chemical residues pose environmental hazards and health concerns. Thus microbial insecticides are seen as an alternative means of pest control that can play an important role in integrated pest management systems and reduce our dependence on chemical pesticides.

2. Microorganisms in Pest Control

Microorganisms that are pathogenic to insects have relatively narrow host ranges which makes it possible to reduce specific pest populations while natural predators and beneficial insects are preserved or given opportunity to become established.

Organisms like viruses, bacteria, fungi nematodes, protozoans and rickettsiae are pathogenic to array of pest species inflicting cultivated plants. Of these, fungi, viruses and bacteria are involved greatly at the present scenario. There are over 1200 species of viruses, fungi, bacteria, protozoa and nematodes known to cause diseases in insects.

Insect resistance to microbial pesticides tends to be less common or to develop more slowly than resistance to chemical pesticides. As far as is known, entomopathogens now produced commercially do not affect humans or animals other than their target species and their by-products are biodegradable.

3. Microbial Pesticides

A microbial insecticide is expected to be

- Convenient to apply (as a dust, spray or bait alone or in conjunction with chemicals, predators, parasites or other pathogens)
- Storable
- Economical
- Easy to produce
- Virulent against target pest
- Safe and acceptable

Microbial insecticides are currently marketed under a variety of names such as biopesticides, biocides, bacterial insecticides, mycoinsecticides, viral insecticides, wettable powders of entomophilic nematodes etc (Moore and Prior, 1993; Soper, 1982).

4. Production and Marketing of Microbial Pesticides in India

Production of microbials in the developed world has concentrated on high technology methods involving complex equipment and sterile culture conditions. For developing countries, research has shown that cheap production systems using local raw materials and equipments can be developed.

As far as India is concerned, Nucleo Polyhedro Viruses (NPVs), bacteria and mycoinsecticides are in use currently in various crops. Production and availability of microbial pesticides is low due to several limitations. Low adaptation by the farmers is mainly due to their slow action (Jayaraj, 2001).

Formulations of *Bacillus thuringiensis* are available commercially with consumption of 15-20 metric tonnes. About 20% of commercial products of microbials (*B. thuringiensis, Beauveria bassiana, Metarrhizium anisopliae*) are now available. Similarly commercial formulations of Nucleo Polyhedro Viruses (NPVs) against *Helicoverpa armigera* and *Spodoptera litura* are available in India. Microbials are expected to replace atleast 10% of chemical pesticides which means a market share of Rs. 60 Crores annually (Dubey *et al.*, 2001).

The Department of Biotechnology (DBT), Government of India has provided financial support to establish units for mass production of *Chrysopa* (Predator), *Trichogramma* (Parasite) and NPV of *Spodoptera* and *Helicoverpa* (target pests) at state agricultural universities and Indian Council of Agricultural Research Institutes besides certain private entrepreneurs (Crop health, Mahated Group, AgroEvo India Limited, BCRL, etc.) are involved in the production of microbials.

5. Advantages of Microbial Pesticides (Narayanasamy, 1995)

- High degree of specificity
- No adverse effect on non-target organisms
- Absence of residue build-up in the environment
- Compatible with chemical pesticides
- They aid growth throughout and are easy to apply
- Relatively cheaper by 50% as compared to chemical pesticides

6. Engineering Microorganisms for Pest Control

The genetic engineering of a microorganism requires three components: a gene that encodes an insecticide, a plant colonizing the microbe and the technology to transfer and express the gene in the recipient colonizing microbe.

Biotechnology is concentrated on microorganism especially bacteria, viruses, protozoa and nematodes. In bacteria, efforts are onto increase the host range, by combining different genetic forms of the d - endotoxin which have virulence within and between insect groups.

In fungi and viruses, efforts are on to increase virulence. Development of virus resistant plants by incorporation of viral Deoxyribo Nucleic Acid (DNA) into crops has offered exciting prospects. This has been achieved in a range of vegetables by incorporation of genes for viral coat proteins into crop genomes.

Cowpea trypsin inhibitors have been engineered into tobacco and tomato. A genetically modified stain of the bacterium *Agrobacterium radiobacter* was the first genetically manipulated organism to be released for pest management. Inoculation of this bacterium protects a number of crops from infection by the crown gall bacterium, *Agrobacterium tumefaciens*.

Recombinant DNA technology and Genetic engineering have enabled the production of several transgenic crop varieties with improved agronomic characters and also insect resistance properties.

At present these are 28 varieties of genetically engineered plants have been commercialized in USA, South America, Japan and Australia.

7. Researchable Areas on Microbial Pesticides

There is ample scope for microbials in the pesticide world; however, their popularity needs attention. There are a number of researchable issues that need attention, as listed below.

- Improvement in shelf life and toxicity
- Consistency in concentration and dosages
- Phytotoxicity resulting from repeated applications
- Safety to non-target organisms
- Effects on endocrine system of larger animals
- Genetic changes in the host
- Standardized bioassay procedures
- Adverse effects to humans during production and use

Among the insect pathogens, fungi are observed to be predominant and potential against insect pests especially those of rice crop as the required humidity and temperature are very much available. The very same ecosystem enables the attack of wide ranging categories of pest species. An account of the humid-based field situation, occurrence of innumerous forms of mycoses of insects in rice has become common and steady. Investigation of field occurrence of such fungi in rice has projected an array of species viz., *Pandora delphacis, Metarrhizium flaroviride* var. *minus, Mucor hiemalis, Rhizopus orzae, Nomuraea rileyi, Beauveria bassiana, Hirsutella* sp and *Fusarium* sp.

Fungi of the late, have become an important component of the biological pest management strategy. As for the natural control features taking place in the rice field , fungal pathogens have been known to be predominant and potential in diseasing an array of pests like brown planthopper (BPH), green leafhopper (GLH), stem borer, leaf folder, grasshopper, black bug etc. Though infection of both adult and immature stages such as caterpillar, nymphs of leafhoppers and planthoppers, beetles, bugs, stem borer, leaf folders etc. was a common feature, cases of adult moths of Lepidopteran pests succumbing to fungal onslaught were encountered by us for the first time in India (Udayaprabhakar, 1994; Narayanasamy, 1994; Narayanasamy, 2001)

- *Pandora delphacis* showed its potentials to kill both brown planthopper (BPH) and green leafhopper (GLH) in Tamil Nadu (Narayanasamy *et al.*, 1992).
- *Zoophthora radicans* had caused large scale casualities of leaf folder in field situation.
- *N. rileyi* was identified to be a virulent pathogen of cut worm.
- *B. bassiana* had wide host range including rice weevil, scarabaeid beetle, GLH, BPH coccinellid beetle and tipulid fly.
- *Aspergillus terreus* was recorded infecting adults and nymphs of the grasshopper *(Hieroglyphus banian).*
- New record of fungi infecting adult moths of leaf folder, stem borer, case worm and cut worm included *Cladosporium* sp. *Mucor hiemalis, Fusarium moniliforme* and *Scopulariopsis* sp.
- Despite unearthing of umpteen fungal species from the rice ecosystem, formulation of then as Mycoinsecticide has not been much worked out.
- Mycoinsecticide, 70% wettable powder, formulation of *P. delphacis* was developed with fungal spores, sorghum grains, clay and teepol and evaluated against BPH and GLH in Tamil nadu and Karnataka.
- Results from Coimbatore and Mandya indicated that highest mortality of 87.8% of BPH was due to the mycoinsecticide sprayed twice at 30 and 50 DAT. Similar result was obtained at Annamalai Nagar with 66.15% and 64.11% kill of BPH (both nymphs and adults) when compared to mycoinsecticide 10%. Dust and BHC 10% dust. The field from where the evaluation of the fungal pesticide was done, epizootics of *P. delphacis* on the BPH and the GLH with 76%-80%, were witnessed after a season.

- From these, it is observed that there are plenty of fungal pathogens which are Lepidoptericidal (larvicidal, mothcidal), Homoptericidal, Coleoptericidal and Diptericidal in action. Certain instar-specific fungi isolated show much promise to checkmate the pests concerned at their weak point (stage).
- Utilization of the mothcidal fungi attract our attention for curbing the moth stage in lieu of the larvae which become direct infestors of the rice plant. Development and utilization of myconinsecticide is only in infant stage in India whilst many other countries have a couple of fungal pesticide. This situation warrants concerted efforts to produce Mycoinsecticide for various crop ecosystems besides rice.

During the course of nearly 15 years of research, thirty five groups of fungal diseases infecting major pests of rice like BPH, GLH, leaf folder, stem borer and minor pests like cutworm, skipper, coccinellid and curculionid beetles and mite were isolated from Tamilnadu, Pondicherry and Karnataka. Pathogens like *Pandora delphacis, Mucor hiemalis, Rhizopus oryzae, Metarrhizium flavoviride* var. *minus, Metarrhizium* sp., *Fusarium* sp., *Entomophthora* sp. and *Beauveria bassiana* infected BPH. Specifically *Pandora delphacis* exercised 61% of mortality of BPH under field condition in the eastern coastal belt of Tamilnadu. In extreme cold weather which fell during November – March in the region, 100% infection of all nymphal instars of the planthopper due to the fungus was witnessed. At times, even dryniid-parasitized planthoppers were not spared. Green leafhoppers were infected with *Pandora delphacis, Metarrhizium flavoviride, Fusarium* sp., *Entomophthora aulicae* and *Beauveria bassiana. Zoophthora radicans* effected 84% infection of rice folder larvae under field condition. The leaffolder larvae were infected with fungi like *Zoophthora radicans, Z. phytonomi, Metarrhizium* sp. *and Rhizopus* sp. *Nomuraea rileyi* proved itself to be a potential pathogen of rice cutworm larvae with 90-92% parasitism, observed in an outbreak of the pest. Occurrence of *Beauveria bassiana* was also detected on various insects like larvae of stem borer, leaffolder, beetles, tipulid flies.

Infections due to certain fungal pathogens on the moths of major pests of rice have been recorded substantially from the Eastern Coastal belt of Tamilnadu. *Cladosporium* sp. was isolated for the first time in India from the case worm and subsequently few more species like *Mucor hiemalis* Wehmer, *Fusarium moniliformae* (Sheld) and

Scopulariopsis sp were recovered from the bodies of yellow rice borer, leaf folder and yellow hairy caterpillar while *Scopulariopsis* sp was gathered from rice cut worm.

Four forms of Mycoinsecticides as 10% Dust and 70% Wettable powder were developed with *Pandora delphacis* (Hori) Humber, a potential pathogen of brown planthopper and green leafhopper on sorghum grains as carrier plus other ingredients (Udayaprabhakar, 1995; Narayanasamy, 1993). The fungal biopesticide, Mycoinsecticide 70% WP comprising of sorghum grains, clay and teepol applied at 2.5 kg/ ha showed its supremeness against BPH with 66.15% and 64.11% mortality respectively at two locations in Cuddalore District, Tamilnadu. Similarly the biopesticide was found to be efficient against the BPH at other locations like Coimbatore, Mandya and Hyderabad.

Cost of one kilogram of the product works out to just Rs 10/-. The pathogen was found absolutely safe to non-target organisms like albino rat, frog, fish earthworm, honey bee, silk worm etc.

Field application of certain mothcidal fungi had showed efficiency in the infection by *Mucor hiemalis, Furasium moniliformae* on stem borer and leaf folder.

Further, another new strategy evolved was that azygospores of *Zoophthora radicans* were mass produced and tested in the field against leaf folder. This has brought out possible curbing of the leaf folder larvae significantly.

From these it becomes obvious that application of the Mycoinsecticide has great scope for combating major pests of rice including brown planthopper and leaf folder and the practice will hold good as an appropriate alternative to chemical insecticides to usher in production of organic crop produce.

8. Conclusions

There is a lot of scope for microbial pesticides as an effective alternative to toxic synthetic pesticides. The methods of production and application technology require attention considering the locally available raw materials and equipments.

Quality control is an important aspect for the ultimate success of microbial pesticides.

It is in this context, that microbial control of pests in agriculture will play an ever increasing role in the development of stable pest management strategies without causing human health and environmental concerns.

Acknowledgement

The authour expresses his gratitude to the authorities of Annamalai University for providing necessary facilities and permission for the work.

References

Moore, A. and Prior, C. 1993. The potential of Mycoinsecticides, *Biocont. News Inf.* 14(2): 31N-40N.

Narayanasamy, P. 1994. Development and use of mycoinsecticide from indigenous fungal pathogen against the brown planthopper (*Nilaparavata lugens* (Stal) problem in rice, Final Project Report, Ministry of Environment and Forestry, Govt. of India, pp.95.

Narayanasamy, P. 1995. Mycoinsecticide: A novel biopesticide in Indian scenario, *Biotech. Develop. Rev.* pp. 15-20.

Narayanasamy, P. Udayaprabakar, L. and Humber, R.A. 1992. *Pandora delphacis,* (Hori.) Humber, pathogen of brown planthopper and green leafhopper in India, *FAO Pl. Prot. Bull.* 40(3): 111-112.

Narayanasamy, P. 2001. Indigenous pest suppression pp.87-112. In: Upathyay, R.K., K.G. Mukerji and B.P. Chamola (Eds). Biocontrol potanital and its exploitation sustainable agriculture. Vol.2: Insect pests Kluwer Academic / Plenum Publishes, New York. 421p.

Soper, R.S. 1982. Commercial mycoinsecticides, In: *Proc. 3rd International Colloquium on Invertebrate Pathology,* Bringhton, pp.98-102.

Udayaprabakar, L. 1994. Investigations on collection identification and formulation of *Pandro delphacis* (Hori) Humber as mycoinsecticides against rice brown planthopper, *Nilaparvata lugens* (Stal), Ph.D. Thesis, Annamalai University, Annamalainagar, India. pp. 306.

❑❑❑

SECTION - IV

BOTANICAL INSECTICIDES

19

Potential of Botanical Biopesticides for Efficient Pest Management Strategies

OPENDER KOUL
Director, Insect Biopesticide Research Centre
30 Parkash Nagar, Jalandhar 144003, India

1. Introduction

Plants are biochemists *par excellence*. During their long evolution plants have synthesized a diverse array of chemicals to prevent their colonization by insects and other herbivores. It is estimated that there are about 2,50,000 to 5,00,000 different plant species in the world today (Dhaliwal and Koul, 2007). Only 10% of these have been examined chemically indicating that there is enormous scope for further work. Over the years, more than 6000 species of plants have been screened and more than 2500 plant species belonging to 235 families (Saxena, 1998) were found to possess biological activity against various categories of pests (Table 1). This number seems to be far less than the actual number of naturally occurring pesticidal plants as this is just 0.77% of the total 3,08,000 species of plants or 0.87% of 2,75,000 species of flowering plants (Singh, 2000). It is thus likely that novel and potent molecules that can be used for pest suppression remain to be discovered from many plant species.

Plants are known to produce a diverse range of secondary metabolites such as terpenoids, alkaloids, polyacetylenes, flavonoids, unusual amino acids, sugars, etc. The structures of more than 600 alkaloids, 3000 terpenoids, several thousands of phenylpropanoids, 1000 flavonoids, 500 quinones, 650 polyacetylenes, and 4000 amino acids have already been elucidated (Metcalf and Metcalf, 1992). Many of these chemicals protect the plants from pests and pathogens. The 10 most important plant families containing the pesticidal plants are listed in Table 2. Most of the pesticidal plants occur in Meliaceae, which has more than 500 species. Some of the insecticidal plants along with their bioactive components and the types of effects produced in target pests are listed in (Table 2).

Table 1: Types of pest control activity reported in various plant species

S.No.	Pest control activity	No. of species of plants	S.No.	Pest control activity	No.of species of plants
1.	Insecticidal	1053	9	Fungicidal	100
2.	Antifeedant	230	10	Bactericidal	4
3.	Repellent	225	11	Molluscidal	6
4.	Growth inhibitor	32	12	Herbicidal	14
5.	Attractant	27	13	Antiseptic	35
6.	Rodenticidal	29	14	Fish poison	147
7.	Acaricidal	2	15	Arrow poison	90
8.	Nematicidal	58	16	Poisonous	69

Source: Grainge and Ahmed (1988)

Table 2: Important plant families having pesticidal plants

S.No.	Family	Number of plants	S.No.	Family	Number of plants
1.	Apocyanaceae	39	6.	Meliaceae	>500
2.	Asteraceae	70	7.	Myrtaceae	72
3.	Euphorbiaceae	63	8.	Ranunculaceae	55
4.	Fabaceae	57	9.	Rosaceae	27
5.	Leguminosae	60	10	Rutaceae	39

Source: Singh *et al.*, (1999)

2. Traditional Botanicals

The commercial use of plant extracts as insecticides began in the nineteenth century with the introduction of nicotine from *Nicotiana tabacum,* rotenone from *Lonchocarpus* sp, derris dust from *Derris elliptica* and pyrethrum from *Chrysanthemum cinerariaefolium.* Rotenones, the first generation botanical pesticides have been extensively used in the past to control household and agricultural pests. Its use, however, had to be dispensed with due to high fish and/or mammalian toxicity. Nicotine, an alkaloid obtained from *Nicotiana tabacum, N. rustica* and *N. glutinosa* is another well-established botanical insecticide. Nicotine analogues like nor-nicotine and anabasine also possess insecticidal properties. Nicotine is active against piercing-sucking insects such as aphids, leafhoppers, whiteflies, thrips and mites. However, due to high mammalian toxicity and detrimental effect on human health, its use as an insecticide has decreased considerably.

Sabadilla alkaloid derived from Sabadilla (*Schoenocaulon officinale* A. Gray) and number of *Veratrum* species, generally referred to as *Veratrum* alkaloids is also known for its insect control properties. The insecticidal activity of sabadilla comes from the alkaloid fraction, which constitutes 3–6% of the extract. Two most important lipophilic alkaloids in the extract have been identified as veratridine and cevadine, the former being more insecticidal. The major effects of sabadilla poisoning include muscle rigor in mammals and paralysis in insects.

Pyrethrum, the most widely used botanical insecticide is extracted from the flowers of *Chrysanthemum cinerariaefolium* (pyrethrum). It is highly effective against houseflies, mosquitoes, fleas and lice. The toxins namely pyrethrins, cinerins and jasmolins have some unusual insecticidal properties, most striking being the immediate knockdown or paralysis on contact which causes most flying insects to drop almost immediately upon exposure (Casida and Quistad, 1994).

3. Isobutylamides

A large number of unsaturated isobutylamides have been isolated from various species of genus *Piper* (Piperaceae), which are known to have diverse insecticidal actions. The compounds have been isolated from the fruits, stem and leaves of various *Piper* species such as *P. nigrum, P. acutisleginum, P. khasiana, P. longum, P. pedicellosum* and *P. thomsoni* (Parmar and Walia, 2001). Screening of other species in the genus points to numerous other potential sources of natural insecticides, such as *P. retrofractum* from Thailand, *P. guineense* from West Africa and *P. tuberculatum* from Central America (Isman, 2001).

4. Glycolipds and Sugar Esters

Recent research in plant resistance to insect pests of *Nicotiana spp., Solanum spp.* And other plants have demonstrated that leaf exudates are effective and prevent plants from any pest attack.

5. Essential Oils and Volatiles

Plant essential oils have also been described as potential source of insecticides that besides being soft and environmentally safe are active against insect pests of agricultural, public health and veterinary importance. The volatile constituents of essential oils mainly include terpenes, hydrocarbons, phenols and other miscellaneous compounds such as alcohols, ketones, aldehydes, carboxylic acids and oxides. Eugenol from *Eugenia cryophyllus*l, 1,8-cineole from *Eucalyptus globules*, citronellal from lemon grass *Cymbopogon nardus*, pulegone from *Mentha pulegium*, and thymol and carvacrol from *Thymus vulgaris* and limonene and linalool from citrus peels are some of the most active essential oil constituents having significant pest control properties.

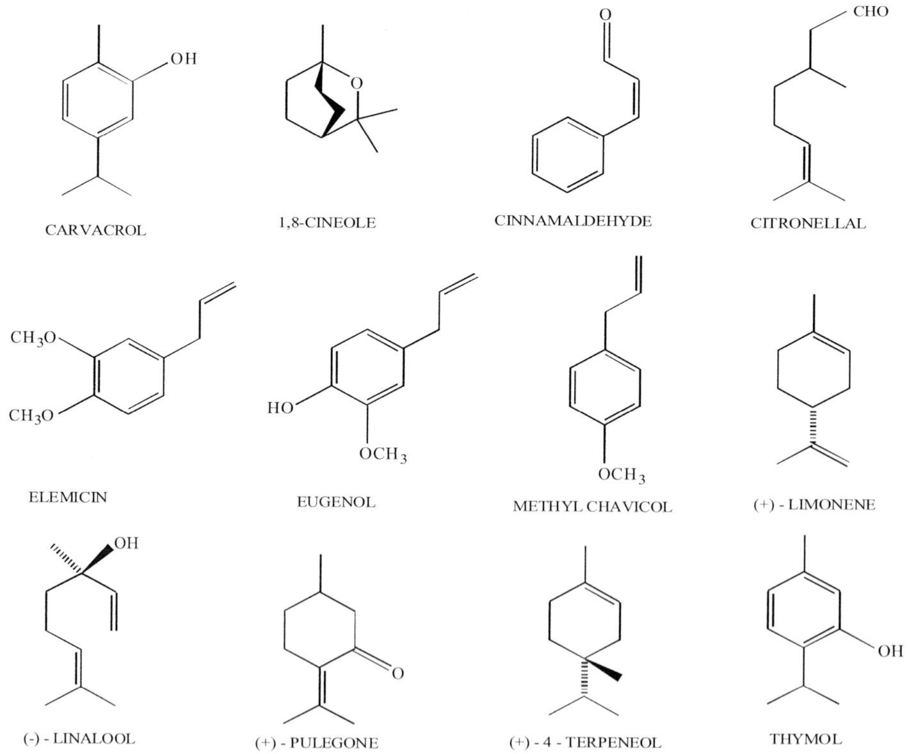

A general structure of a sugar ester

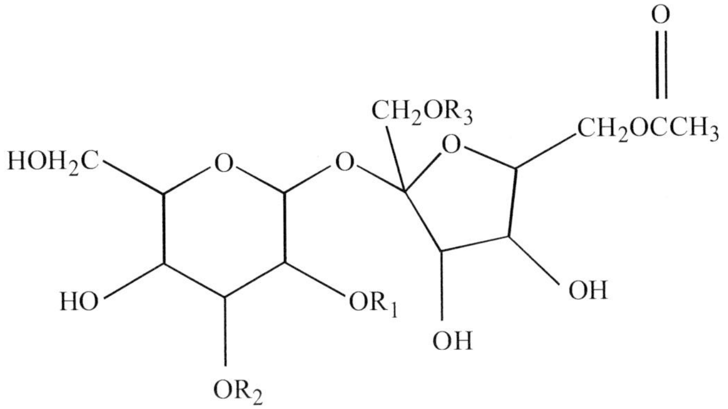

A general structure of a sugar ester

The turmeric leaves, the unutilized part of turmeric plant, on hydrodistillation yielded oil rich in a-phellandrene (70%) that inhibited growth of *Spilosoma rabica* and caused larval mortality at 1% concentration. Turmeric rhizome oil and its major constituent turmerones on the other hand were less active. The leaf oil also showed ovicidal and nymphicidal action against *D. koenigi* and exhibited moderate knockdown effect against *Tribolium castaneum*. Curcumene and ginger oil obtained from *Zingiber officinale* rhizomes at 0.2% concentration showed 86% inhibition of the mycelial growth of the test fungus *R. solani*. Some dillapiole derivatives show ovicidal action against the yellow stem borer *Scirpophaga incertulas* eggs (Vidyasagar and Walia, 1987). The semi synthetic derivatives of dillapiole namely heptanoyl dihydrodillapiole and dillapiole dichlorophenyl cyclopropane inhibited egg hatching and were effective even against older egg masses.

6. Limonoids

Neem (*Azadirachta indica*) and chinaberry (*Melia azedarach*) trees belonging to Meliaceae are a storehouse of large number of pesticidally active limonoids. Although neem seeds and leaves have been used as traditional insecticides for centuries, modern interest in neem as an insecticide dates back to the work of Pradhan *et al.* (1962) who reported that dilute aqueous kernel extracts prevented feeding of the desert locust, *Schistocerca gregaria*. Of the many active bitter principles isolated from the neem, azadirachtin is the most potent. Among the large number of tetranortriterpenoids isolated from seeds, leaves and bark, azadirachtins, salannin and nimbin. Exhibit significant biological activities and revealed interesting structure activity relationships. In addition to being a remarkable antifeedant, azadirachtin interferes strongly with molting and reproduction in several species of insects. While azadirachtin A is present in the extract to the extent of 85%, azadirachtin B and H occur up to 15% only. Other azadirachtin congeners occurred as minor constituents in neem seed extracts. Very recently a fast and economic reverse phase medium-pressure liquid chromatographic method has been reported for the isolation and characterization of three major azadirachtins A, B and H in adequate quantities from seed kernel extractives. While all three azadirachtins exhibited nematicidal and antifungal activities, azadirachtin B was the most effective against the reniform nematode *Rotylenchulus reniformis*. Studies conducted on the effect of azadirachtin and aqueous neem seed kernel concentrate on germination and seedling growth of Soybean and Maize revealed that neem extractives exhibited significant phytotoxicity. In another study, neem extractives have been found to strongly inhibit germination and growth of several crops such as bean, carrot, raddish, rice and sesame; and weeds such as *Echinochloa*.

Melia azedarach, and *M. toosendan* are two other Meliaceae plants, which contain several limonoids possessing insect antifeedant activity. From the stem and root bark of *M. azadarach,* several azadirachtin type insect antifeedants have been isolated. Some of these include meliacarpinins, azedarachins, toosendannins, ammorastatins, amoorastatone, trichilins and nimbolidins, etc. Among the various meliacarpinins, 1-cinnamoyl-3-acetyl-11-methoxymeliacarpinin with antifeedant index (AI_{50}) of 50 ppm was the most active.

Citrus seeds, available as waste by-products of the citrus processing industry, are a cheap source of bitter limonoid limonin possessing significant anti-insect properties. In addition to having antifeedant action against various insects, limonin induces antifeedant action against insect pests, which have developed resistance to conventional synthetic insecticides. About eleven limonoids have been identified from various citrus seeds of which obacunone, limonin and nomilin have been reported as potent antifeedants against the termite *R. speratus*.

7. Rocaglamides

Among the various botanicals isolated from *Aglaia odorata, A. elliptica,* and A. *duppereana* (Meliaceae), rocaglamide is the most effective (EC_{50} = 0.8 ppm). It is slightly more potent than azadirachtin

(EC_{50} = 1.0 ppm) against some insect species (Janprasert *et al* 1993). As growth inhibitors, rocaglamide and methyl rocaglate are almost similar in their activity (EC_{50} = 0.9 ppm) and quite comparable to azadirachtin (0.26 ppm) and so are the aglaroxins isolated from other *Aglaia* species (Koul *et al.*, 2005a,b).

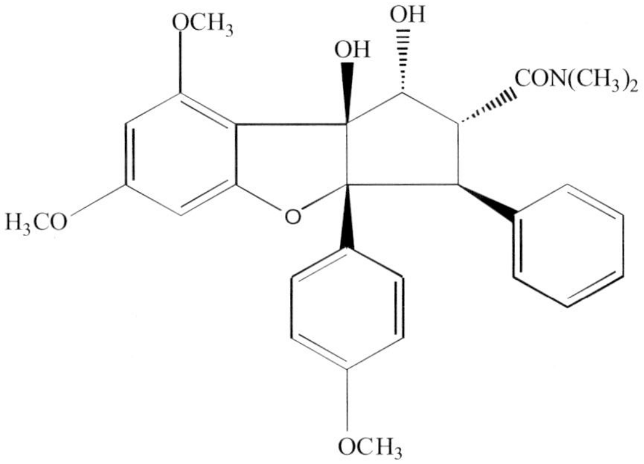

Rocaglamide

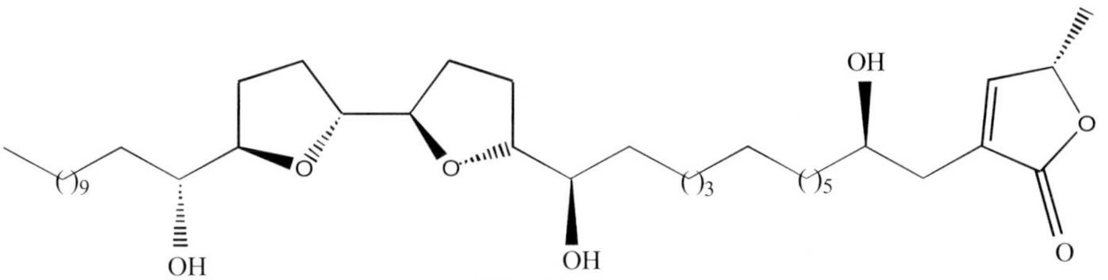

ASIMICIN

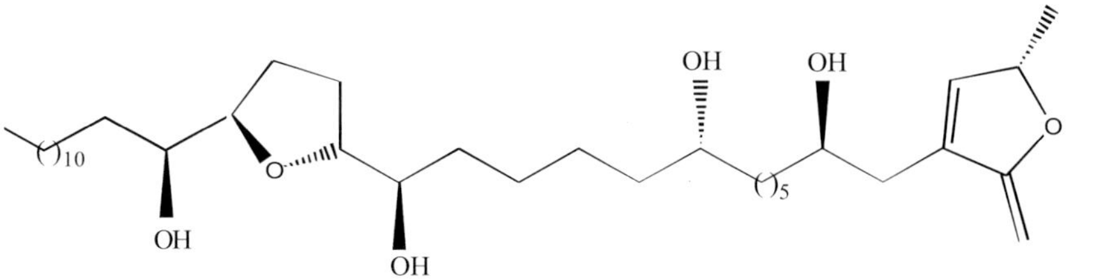

ANNONACIN

Structure of some acetogenins

8. Acetogenins

Bioactive acetogenins such as annonins, and related compounds namely squamocin, asimicin, annonacins and cohibinsin occur widely in twigs and branches, unripe fruits and seeds of several *Annona* (custard apple) species (Annonaceae). Entire group of annonaceous acetogenins has been patented as pesticide in which asimicin was claimed as a structurally defined pesticidal acetogenin. Johnson *et al.* (2000) have isolated hundreds of acetogenins from the Annonaceae, and for many their potential as anticancer agents exceeds their value as insecticides. According to Isman (2006) Annona seed extracts may prove useful in tropical countries where the fruits are commonly consumed or used to produce fruit juice, in which case the seeds are a waste product.

9. Light Activated Allelochemicals

Ultra violent and sunlight usually play a counter-productive role in degradation of botanical pesticides leading to decrease in their effectiveness. In some cases, toxicity of some of the phytochemicals increased following their exposure to light radiations. Such light-activated phototoxins like substituted acetylenes, thiophenes, acetylenic thiophenes, quinines, furanocoumarins and related compounds exhibit significant pest control properties.

10. Miscellaneous Compounds

Phytopesticides that have not been commercialized despite being active, include pipericide from *Piper nigrum*, quassin from *Quasia* species, and pellitorine from *Anacyclus pyrethrum*. A new group of plant-derived napthaquinones from the Chilean plant *Calceolaria andina* reportedly cause rapid knockdown and kill of flying insects but are too unstable for use as products themselves. Turmeric (*Curcuma longa*) is known for its colouring, flavouring, digestive, healing and antioxidant properties. Its extractives and constituents are known for their stored grain protection and insect repellent action. Very recently detailed studies have been conducted to assess pest control potential of phytochemicals derived from turmeric and ginger (Agarwal and Walia, 2004). The activity of curcumin derivatives, turmeric oil and turmerone was comparable to commercial neem formulation. Azadirachtin-turmeric oil combination has been found to be more effective than azadirachtin or turmeric alone against the Bihar hairy caterpillar *Spilosoma rabica*. Ar-Turmerone, the major constituent of turmeric rhizome oil and labda-8(17),12-diene-15,16-dial exhibited considerable mosquitocidal activity. Ginger oleoresin extracted from fresh rhizomes of *Zingiber officinale* is a complex mixture of several closely related phenolic alkanones such as gingerols, shogaols, gingerones, paradols, gingerdiols and diarylheptanoids. These compounds possess varied degree of pest control properties. Ginger based products have also been found to exhibit moderate insect growth regulatory and antifeedant activity against *Spilosoma rabica* and significant antifungal activity against *Rhizoctonia solani.*

11. Insecticide Synergists of Plant Origin

Botanical pesticides usually degrade rapidly or detoxified by enzymes inside the target pests. Because of their rapid breakdown, the insecticide is not capable of providing the desired effect. Accordingly, synergists have been added to avoid detoxification process in insects and enhance insecticidal action of the products. High cost insecticide synergists, pyrethrins and several other natural insecticides are often mixed with the commercial insecticide synergists piperonyl butoxide (PBO) to enhance their effectiveness.

12. Botanicals and Integrated Pest Management

Botanical pesticides can be integrated with the use of parasitoids and predators, microbial pesticides and even synthetic chemical pesticides to achieve greater efficiency in pest control. Several neem derivatives and their formulations have been evaluated against several parasitoids and predators with variable results. The application of neem seed rabic extract (NSKE) for the control of *S. litura* did not affect the emergence of the egg parasitoid, *Telenomus remus* Nixon, and predator, *Brinckochrysa scelestes* (Banks).

The plant extracts of some rice varieties were found to be highly toxic to important rice pests including *Chilo suppressalis* (Walker), *Nilaparvata lugens* (Stal) and *Sogatella furcifera* (Horvath) but were safe to predators of hopper complex, viz., *L. pseudoannulata* and *C. lividipennis.* In the rice field trials, application of neem seed bitters as ULV sprays has no detrimental effect on a number of natural enemies including the black cricket predator, *Mettioche vittaticollis;* ground beetle predator, *Ophionea nigrofasciata* (Schmidt-Goebel) and lady beetle predator, *Conocephalus longipennis* (deHann). Predation of leaf folder, *Marasmia patnalis* Bradley eggs by *M. vittaticollis* was unaffected in treated fields. The parasitization of *Cnaphalocrocis medinalis* (Guenee) larvae was relatively unaffected in rice fields sprayed with 50% neem oil (ULV application) than in control fields. Neem oil 50% sprayed at rates up to 20 ml of spray fluid (recommended rate 3 ml/litre) has no adverse effect on *Chrysoperla carnea* (Stephens) and *Tetrastichus howardi.*

Botanical pesticides are known to reduce the population of pests to the manageable level from where the natural enemies take control of the pest population. Because of the low residual action, the release of natural enemies can be coupled with the spraying schedules as the mortality of the natural enemies in the treated plants is generally negligible. Treating the eggs with margosum (1500 ppm) anytime after parasitization by *Trichogramma chilonis* Ishii did not affect hatching of *T. chilonis* and was safe to *Bracon brevicornis* Wesmael, *Lycosa pseudoannulata* Bosenberg & Strand, and *Cheilomenes sexmaculata* (Fabricius). Combination of Nimbecidine (2%) along with other control measures was recommended for the management of *Helicoverpa armigera* (Hubner) on cotton.

Botanical pesticides do show rabicapids with synthetic chemical insecticides. The use of neem formulations in combination with conventional insecticides resulted in better control of pink bollworm and spotted bollworm on cotton. The use of neem pesticides and alphamethrin on cotton proved as good as the sprays of alphamethrin alone in checking the menace of bollworms and increasing the seed cotton yield.

A combination spray of NeemAzal T/S 1% (0.5 l/ha) and monocrotophos 36SL (1.25 l/ha) resulted in highest reduction of damage due to rice stem borer, *Scirpophaga incertulas* (Walker).

13. Current Trends in the Use of Botanicals

In United States the botanicals registerd for use are pyrethrum, neem, rotenone, several essential oils, sabadilla, ryania, and nicotine. Several azadirachtin based insecticides are sold in USA and number of essential oils are exempt from registration. Canada has been more conservative where only pyrethrum, rotenone, nicotine and few essential oils are registered for use. Mexico, of course allows the products registered in USA.

In Europe pyrethrum, neem, rotenone, and nicotine are allowed, however, there is variation in registration protocols in different European countries. In fact, neem has still to make headway in these countries.

In Asia, India leads in the use of botanicals where number of products is registered under provisional registration. Neem based products are in abundance in addition to pyrethrum, rotenone, nicotine and essential oils. However, neem is yet to be approved in Australia, New Zealand, and Philippines.

In Latin America, Brazil leads in the registered products based on pyrethrum, rotenone, neem, garlic and nicotine. Throughout Latin America plant oils and extracts are produced by cottage industry. However, data on regulated products for most of African countries is not known. Apparently only pyrethrum is approved for use in South Africa (Isman, 2006).

Overall there appear to be three barriers to commercialization for botanical biopesticides, i.e. sustainability of the botanical resource; standardization of complex extracts and regulatory approval. However, real benefits of botanicals can be realized in developing countries where traditional use of plants and plant derivatives is long established, particularly for protection of stored products.

References

Casida, J. E. and G. B. Quistad, 1994. *Pyrethrum flowers, producton, chemistry, toxicology and uses.* Oxford University Press, Oxford.

Dhaliwal, G. S. and O. Koul, 2007. *Biopesticides and Pest Management: Conventional and Biotechnological Approaches.* Kalyani Publishers, New Delhi.

Isman, M.B. 2001. Biopesticides based on phytochemicals. pp. 1-12. In: O. Koul and G.S. Dhaliwal (Eds) *Phytochemical biopesticides.* Harwood Academic Publishers, Amsterdam, The Netherlands.

Isman, M.B. 2006. Botanical insecticides, deterrents and repellents in modern agriculture and an increasingly regulated world. *Annual Review of Entomology,* 51: 45-66.

Johnson, H. A., N. H. Oberlies, F. Q. Alali, and J. E. Mc Laughlin, 2000. Thwarting resistance: Anonaceous acetogenins as new pesticidal and antitumor agents, pp. 173–183. In: S. J. Cutler and J. G. Cutler (Eds.), *Biological active natural products: Pharmaceuticals,* CRC Press, Boca Raton, FL.

Metcalf, R. L. and E. R. Metcalf, 1992. *Plant kairomones in insect ecology and control.* Chapman & Hall, New York.

Pradhan, S., M.G. Jotwani and B. K. Rai, 1962. The neem seed deterrent to locusts. *Indian Farming,* 12: 7–11.

Singh, R. P. 2000. Botanicals in pest management: An ecological perspective, pp. 279–343. In: G. S. Dhaliwal and B. Singh (Eds.), *Pesticides and environment.* Commonwealth Publishers, New Delhi.

❏❏❏

20

Secondary Plant Compounds in Insect Pest Management Practical Experiences and Prospects

ARIVUDAINAMBI, S.
Reader, Department of Entomology, Faculty of Agriculture
Annamalai University

1. Introduction

Ecologically active phytochemicals are elaborated as secondary metabolites or secondary plant compounds. The secondary metabolites have more limited distribution among plants and frequently have little obvious survival value out side and ecological context. Although the physiological role of many secondary plant compounds is still obscure, their ecological function is currently well recognized. They are non-nutrients and qualitatively much different in their chemistry. It is estimated that more than 5,00,000 Secondary plant compounds occur naturally in plant kingdom.

Keeping in view the ecological significance of secondary plant metabolites, they may be defined as "Non-nutrients produced by one organism that affects the behaviour, health or ecological welfare of another organism of the same or other species". Such compounds are also called as allelochemicals. Allelochemicals are allomones if their activity favours the producing organisms and they are kairomones if their activity favours the receiving organisms. These secondary metabolites of plants act as efficient insecticide against control of harmful pests without causing environmental degradation.

2. Botanical Pest Control

Employment of plant toxic principles in insect control probably originated during prehistoric times. The ancient texts such as Upavana Vinoda; sarangadhara (3th century A.D), Brihat Samhitha Varahamihira (6th century, 500 A.D), Krishi Sukthi; Kashyapa (8-9 th century A.D),Amarakosha; Pandit Amarasimha (6th century A.D), Krishi parasara; Parasara (950-1100 A.D), Dhanvantri Nighantu (11th century A.D) and Raaja Nighantu; Pandit Narahari (11th century A.D) provide more information on agriculture and pest control.

The ground seeds of *Schoenocaulon officinale*, which contain veratrine alkaloids as loose powders, were used by natives of Central America. Rotenoids from Lonchocarpus and Derris have been used as fish poisons by natives of Amazonia and Malaysia for hundreds of years and ground roots and stems were recommended against leaf eating caterpillars in 1849. The tobacco dusts containing nicotine was used as insecticides and was recommended to control the Aphids as early as 1750. The pyrethroids from chrysanthemum were reported as traveller's companions to control fleas and bedbugs before 1800.

The flowers of *Chrysanthemum cinerariaefolium* contains an active principle which is commonly employed to control the fleas. The plants are commercially grown in Kenya, Tanzania, Rwanda and Zaire. The world production is estimated at about 200 metric tonnes of active ingredients with 63% and 22.8% contributed by Kenya and Tanzania. The natural pyrethroids are used as insecticides in three major ways such as a) the rapid "knockdown" agents in houshold sprays and aerosols, usually at 0.05 to 0.2% pyrethrins plus 5 to 10 fold synergist such as b) as livestock and cattle sprays and c) sprays or dusts for mills and warehouses, dusts for protection of stored grains at 0.05% pyrethrins and 0.8% piperonyl butoxide.

The flowers of *cinerariaefolium* contains an average of 1.3% pyrethrins which are concentrated as achenes comprising of about 90% pyrethrins. The active principles are readily extracted with solvents such as petroleum ether, methyl alcohol, or acetone and the concentrate can be freed from waxes and purified to 90 to 100% by extration with nitro methane and adsorption on active carbon.

The chips of two trees picrasma exceisa of Jamaica and *Quarassia amara* of Surinam have been used as insecticides since about 1850. The teas made by steeping the chips in water are effective contact insecticides against aphids and sawflies. The toxic principle found in the chips is the amaroid quassin.

The root and stem of *Ryania speciosa* are dried and ground well which contain about 0.1% to 0.2% of ryanodine alkaloid that has been commercially used to control the European corn borer and other chewing insects.

The veratrine alkaloids of the roots of *Veratrum alba* and *Veratrum* viride have been used as insecticides for more than 100 years but are now of little importance. The veratrine alkaloids are present in ground seeds to about 2 to 4%.

Rotenoids occur in the roots and seeds of more than 68 species of plants including 21 species of *Tephrosia*, 12 of *Derris*, 12 of *Lonchocarpus*, 10 of *Milletia* and several of *Mundulea*. The important varieties for the production of insecticides are *Derris elliptica* and *Derris malaccensis* from Malaysia and *Lonchocarpus utilis* and *Lonchocarpus urucu* from South America. Rotenoids are used as insecticides a) as dusts and water dispersible powders of ground roots used to control a variety of chewing insects prior to harvest and b) as extracts of ground roots emulsified or in ointments for the control of cattle grubs of ox warbles encysted in the backs of cattle. Rotenoid containing roots have 5 to 9% rotenoid and upto 31% ether extractives in Derris elliptical and 8 to 11% rotenone and 25% extractives in Lonchocarpus utilis.

The leaves and stems of *Nicotiana tabaccum* and *Nicotiana rustica* bear relatively simple alakaloids called nicotine which have insecticidal importance comprising about 97% of the total alkaloids. The nicotine can be readily extracted from ground toabacco by treatment with alkali followed by steam distillation or by organic solvents. It is used insecticidally as the free base standardized at 40% or al less volatile and safer nicotine sulfate standardized at 40% nicotine is used as a spray for control of Aphids and other soft bodied insects and sometimes as a dust.

3. Categories of Secondary Metabolites

The secondary metabolites commonly found in plants are Toxic amino acids, Proteinase inhibitors, Alkaloids, Cynogenic glycosides, Phenols, Tannins, Lignins, Flavonoids, Glucosinolates, Toxic lipids, Terpenoids, Saponins and Phytohaemagglutinins.

Toxic amino acids : Among the toxic amino acids, canavanina found in *Cnavalia ensifonnis* caused larval mortality and reduced larval development in *Manduca Sexta.* While L. azetidine-2-carboxylic acid found in *Convallaria majalis* reduced larval development in *Heliothis zea* (Boddie).

Protease inhibitors : Several protease inhibitors have been reported to cause effect on growth and development of different phytophagous insects. For example Trypsin inhibitor -I found in *Vigna ungiculata* caused larval mortality in *Callosobruchus maculants* F. while the seame typsin inhibitor affected the digestive physiology of *Spodoptera exigua* (Hub.) and *H. zea* when found in *Glycine max* and *Solanum tuebrosum.* Similarly, Trypsin inhibitor-II found in *S. tuberosum* affected digestive physiology of *H. zea* and *S. exigua.*

Phytohaemagglutinins : The phytohaemagglutinin found in *Phaseolus vulgaris* caused larval mortality in *C. maculates* while lectin in *G. max* inhibited larval growth in *M. Sexta.* Similarly the E-type lectin found in *P. vulgaris* caused adverse effect on the digestive physiology of *C. maculates.*

Phenolics acetogenius and tannins : Among phenolics, acetogenins, tannins and the total phenols found in *L. esculentum* inhibited larval growth in *H. zea.* While DIMBOA (2, 4-dihydroxy-7-methoxy (2H)-1,4-benzoxazin-3-(4H)-one and MBOA (6-methoxy-2-benzoxazolinone) among acetogenins found in *Z. mays* reduced larval development of *Ostinia nubilalis.* Tannins occurring in *Gossypium hirsutum* adversely affected larval development of *Heliothis virescens* and *Earias vittella.*

Saponins : Saponins found in *G. max* and cac-precipitable saponin in *Pisum sativum* caused larval mortalities and inhibited larval growth, respectively in *Callosobruchus chinensis* (L.)

Alkaloids : Several alkaloids reported from different plant species had their antibiosis effects in insects feeding in them. For example L-tomatin found in *L. esculentum* caused heavy larval mortality of *H. zea* and demissine found in *Solanum* spp. Inhibited larval growth of *L. decemlineata,* while 2-tridemanine occurs in wild tomato (*L. esculentum*) caused high larval mortality of *M. Sexta.*

Glucosinolates : Among the glucosinolates found in *Brassica spp.,* Sinigrin and sinalbin inhibited growth of *Myzus persicae* S. The AP-isothiocynate found in *Brassica* spp. killed the adults of *Drosophila melanogaster* Meigen.

Terpenoids : Several terpenoids reported from different species of cotton are found to cause antibiosis effects in different species of insect pest of cotton. Gossypol found in *Gossypium* spp. Reduced larval growth and development of *H.virescens* and *E. vittella* while heliocide and hemigossypolone found in *Gossypium hirsutum* caused reduction in larval growth of *H. virescens*.

Flavonoids : Out of several flavonoids reported from different plant species quercetin found in *Gossypium* spp. Has been reported to act as growth inhibitor in *P. gossypiella, H. zea, H.virescens* and *Schizaphis graminum*.

While the same flavonoids when found in *Quercus macrocarpa* inhibited the growth of *Scolytus multistriatus*. Similarly myristicin and morin, found in *Q. macrocarpa* inhibited the growth of *Bombyx mori* L and *H. virescens*. Sesamin & Kobusin found in *Magnolia kobus* adversely affected insect growth.

3.1 Mode of Action

Insecticidal Action of Secondary Metabolites

- Neuromuscular block -Strynchnos , Parthenium
- Nerve membrane sodium channels –Ryania
- Generation of more action potentials- neem oil
- Insect cell replication and protein – neem extracts
- Oxidative phosphorylation and inhibit ATP production – Cleistanthus collinus

Deterrence of Secondary Metabolites

- Antifeedancy - Block sensory inputs from chemo receptors, act on mid gut enzymes. Eg. Neem
- Ovipositional deterrent, Repellent action - Block sensory inputs from chemo receptors. Eg. Neem

Growth Regulation of Secondary Metabolites

Insect growth regulation - Prothoracicotropic hormone (PTTH), Ecdysone and Juvenile hormone

Advantages

- Effective against a wide range
- Minimize pest incidence

- Making natural balance with bio- friends
- Increase biodiversity
- Farm level/village level/cottage industry
- Minimize the lateral entry
- Cost effective
- No harmful residue, resistance & resurgence

3.2 Problems Associated

Edaphic Factors

At present, neem germplasms adaptable to various cotton agroecostystems, especially during drought periods, are not readily available. The content of AZ in plant parts varies considerably, according to growing conditions in each agroclimatic zone and depending upon the extraction methods, techniques of harvest, drying and storage of seeds. Some insect pests and plant diseases attacking neem trees during their development need to be controlled.

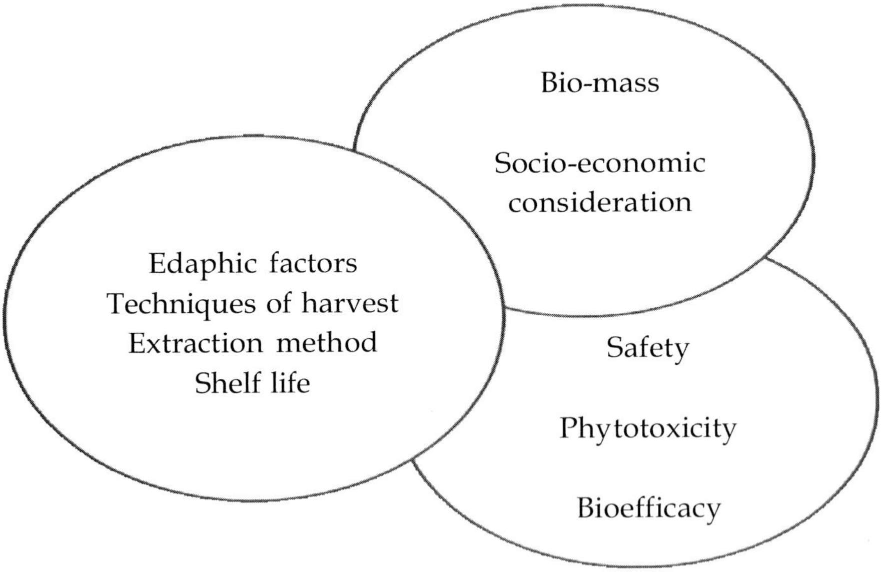

Techniques of Harvest

Active principle concentration of plants varies on the bases of time of collection, plant parts collected, age of the plants.

Extraction Methods

Solvents and methods of extraction play a major role in extracting active principle.

1. Soxhlet extraction method
2. Cold solvent extraction method
   - Crude extraction method
   - Storage

Once kernels are collected in sufficient quantity, they should be well dried and stored in a dry and well-ventilated place. Moist kernels are susceptible to attack by saprophytic bacteria and fungi responsible for biodegradation and thus the quality of neem oil reduced. Also, the fungus *Aspergillus flavus* Link produces aflatoxin. Unfortunately, detection of caracinogenic substances in neem oil has not, to date, been included in any quality control procedure.

Persistence

Any crude preparation must be sprayed as soon as possible because its constituents persist on plants for more than 4 days during which maximum mortality of pests is observed. In the case of S. litura, NSKE (1-10%) effective amongst plant products but caused 100% mortality of fourth instar larvae only 10 days after treatment in the laboratory Likewise, Az and other constituents degrade easily due to chemical changes in the presence of bright sunlight high temperature and organic solvents (Koul *et al.*, 1990:1996). Neem products are also more susceptible to hydrolysis than the organophosphates and carna, ates, and they are degraded easily in water. Therefore, repeated applications are needed for achieving satisfactory pest control. However, the stability of AZ in Nimbecidine during up to 30 months of storage in polythene bags was made possible. Similarly, isolation of hydrogenated derivatives of AZ (e.g. dihydri AZ and tetrahydro-AZ) and radioactive precursors show longer residual activity than the base molecule, and can therefore resist better in sunlight. These analogues may be helpful in improving shelf life and residual quality of neem products meant for field use. To facilitate transport and to avoid rapid degradation, recommended pregelatinized starch encapsulated neem products which are quite stable.

Neem preparations are slow in action, so some damage to cotton is unavoidable. Generally, farmers do not wish to wait after noticing the first appearance of insects or plant damage and prefer to apply synthetics. The commercial products Margocide and Achook were less effective than conventional insecticides against bollworm complex in Punjab (India). Neem oil and formulations were found to be inferior to endosultan (0.06%) in killing S. vittella larvae and Nimbercidine 2500 reduce *H. armigera* only up to 5 days. Similarly, spraying with neem extract (5%) or Neemark (0.5%), Margocide OK or Margocie CK (5- 10 m/l/l) or neem oil (2.5%) could not control sucking pests effectively during vegetative phases of plant growth.

Undesirable Effects on Benefical Organisms

A 2% kernel suspension repelled females of a predatory chrysopid, *Chrysoperla carne* (Stephens) from ovipositing, although no ovicidal was observed when eggs were sprayed.

Inhibition of germination and sporulation of entomopathogenic fungi have been noted. eg. neem oil (5%) on *Metarhizium anisopliae* (Motsch.), Kernel extract or neem oil (2%) on *Beauveria bassiana* (Bals.). Vuill.

NSKE in hexane showed higher toxicity against eggs and pupae of Bracon brevicornis Wesm. Whereas alcohol extract was more toxic against larvae. Neem oil and NSEK are detrimental to the growth of *Nomuraea rileyii* (Farlow), a fungus which attacks Spodoptera spp. Larvae.

Safety

Poisoning of humans due to ingestion of unrefined neem oil and health risks due to handling of neem concentrates and Neem Azal have been reported.

Socio-economic Considerations

Farmers are yet to be convinced of the efficacy of neem in plant protection, and are reluctant to use neem products. Concerted efforts and scheduled programmes are therefore necessary to provide training to farmers, retailers and dealers, and to establish a permanent network for efficient marketing and promotion avenues. Collection of seeds/ kernels and preparation of crude extracts is cumbersome and time consuming, so farmers prefer to purchase synthetic pesticides or ready to use formulations of neem derived products instead of preparing extracts themselves.

Prospects for pest management using botanical products/ pesticides:

Cost

Considering present market prices, neem products are less expensive than other pesticides. Extracts or decoctions or neem leaves, seeds or kernels can easily be prepared by farmers themselves with limited financial resources and labour, reducing the cost substantially. It was emphasized prospects for sustainable use of neem homemade pesticides, although its use depends upon the following factors:

1) availability and cost of raw material; 2) labour requirement; 3) production technology; 4) application rate of raw material; 5) quality and effectiveness of the prepartation; 6) access to and attitude of farmers towards synthetic pesticides; 7) ability and willingness of farmers to devote time; 8) image of neem in the community; 9) plant protection "knowhow" of farmers; 10) awareness of farmers regarding harmful effects of synthetics on human health and environment; 11) utilization of neem by products such as cake as manure.

4. Conclusion

Currently, formulated material is more costly due to emulsifiers, solvents and stabilizers included in it. In order to reduce the manufacturing cost and to facilitate handling and transport, formulations based on dust (2-5%) and seed kernel or oilbased granules (5-10%) are being manufactured. Fresh extracts of leaves and kernel, and oil, may also be advocated for field use.

❑❑❑

21

Utility of Plant Products in Organic Pest Control

VEERAVEL, R.
Professor, Department of Entomology, Faculty of Agriculture
Anamalai University, Annamalai Nagar – 608002.

Plants are biochemists par excellence and during their long co-evolution with insects have biosynthesized a diverse array of chemicals (secondary metabolites) to prevent their colonization by insects and other herbivores. (Dhaliwal & Arora, 2001). Although only 10,000 secondary metabolites have been chemically identified so far, the total number may exceed 4,00,000 and these defensive chemicals belongs to various categories viz., terpenoids, alkaloids, glycosides, polyacetylenes, flavonoids, sugars, amino acids, phenols, tannins etc which cause behavioural and physiological effects on insects.

Plant substances with biological activity can be utilized in three ways

- These active principles can be used directly against insects as crude extracts or are purified and formulated as insecticides on a commercial basis.
- Bio-active plant chemicals often serve as good models for synthetic analogues

- Studies on modes of action and elucidation of chemical structures of bioactive principles may provide leads and insights into the structural activity relationships.

1. History of Plants in Pest Control

The use of plants / botanicals is not new and has been in practice centuries back as ecofriendly pesticides. For thousands of years, people in India placed neem leaves in their books, grainbins, cupboards and closets to keep away insects. Chinese used pyrethrums thousands of years ago to control pests. It has been documented that the use of *Chrysanthemum cinearaerifolium* (Trev.) for its insecticidal properties was known during the period 521-486 BC. *Nicotiana tabacum* L. was being used as a dust or water extract to control insects three hundred years ago. A number of botanicals were used for seed treatment in China, India, Greece and Arabia to prevent insect attack since ancient times.

1200 BC	The Chinese used would ash for pest control in enclosed spaces and pyrethrum for seed treatment.
505-87AD	Varahamithira mentioned the use of root of vasika (*Justicia adathoda* L.) and leaves of atimuktaka (*Hiptage bengalensis*) for seed treatment against insects in India
1000 AD	Surapala in his treatise Vriksaurved, advocates the use of mustard paste, Vidanga (*Emellia ribes*) and lotus stalks
1625-28 AD	Hsu Kuang - Chhi recommended cotton seed oil as an insect repellent
1848 AD	*Derris sp* (rotenone) was reported used in insect control in Asia
1858 AD	Pyrethrum was first used in insect control in USA
1940 AD	Antifeedant activity of leaves of *Melia azedarachta* L. was reported by Volkonsky.
1949 AD	Development of first synthetic pyrethroid, allethrin
1951 AD	Development of carbamate insecticide based on physostigmene an alkaloid from *Bhysostigma venenosum*.
1962 AD	Discovery of antifeedant effect of neem against locust in India.

1966 AD Discovery of juvenile hormone analogues from balsam fir.

1976 AD Discovery of anti juvenile hormone mimic from *Ageratum haustonianum* Mill.

1977 AD Chemosterilant activity of precocenes from *Acorus calamus* L.

1982 AD Juvenile hormone activity of karanjin from *Pongamia glabra*Vent.

1983 AD Fumigant action of essential oils of *Ocimum canum.*

2. Current Scenario of Plant Products

Considerable work has been done in the research and development of botanical insecticides. The main importance has been on field and laboratory evaluation of plant extracts against insects and less on isolation and identification of active principles with few exception. Although about two thousand four hundred plants belonging to 189 plant families (Table 1) have been identified in several countries for their pesticidal properties but only two dozen plants have been found promising.

Table1: Important Plant families having pesticidal plans*

Family	Number of plant species
Apocyanaceae	39
Asteraceae	70
Euphorbiaceae	63
Fabaceae	57
Leguminosae	60
Meliaceae	>500
Myrtaceae	72
Ranunculaceae	55
Rosaceae	27
Rutaceae	39

*Singh *et al.* (1999)

The plant bio-diversity is so wide in India that there are a number of plants, which have properties to be used as pesticides against insect pests. At best, botanical insecticides or plant products, constitute one per cent of the world insecticide market, but annual growth in the range of 10-15 per cent is entirely possible. More and more emphasis is now being given to develop pesticides from naturally occurring plants not only in developing countries with limited options of other control measures, but also attracting attention in developed countries which though food sufficient, are increasingly becoming conscious of food safety.

Pyrethrum ranks first in popularity as an effective natural insecticide, though the synthetic mimics are becoming more popular for their photostability. In India, it is produced in the Nilgri hills in Tamil Nadu. Nicotine is still being used in Europe though it has high mammalian toxicity. On a world - wide basis, about 6,00,000 Kg of nicotine sulfate and 75,000 kg of pure nicotine were produced annually till recently. Rotenone from *Derris sp* was recorded for agricultural use but it has lost popularity due to its photo instability

Among the plant products, neem (*Azadirachta indica* A. Juss) is the only tree that has gained world wide recognition in the last decade. It has been known as medicinal tree of remarkable, manifold curative, including antiinsect properties. In India alone, neem has been evaluated against more than 105 insects, 12 nematodes and a fungi (Singh and Kataria, 1991). It has also been exploited commercially. From 1970 onwards nearly 171 patents have been registered in the world including 36 from India. The potential of azadirachtin lies in its several advantages, including its potency, which is comparable to that of synthetic insecticides. Furthermore, it has a high degree of specificity, affecting behavioural, developmental and biochemical processes peculiar to insects and is non mutagenic, biodegradable and it has synthetic activity in plants, being tranlocated throughout the plant following absorption through the leaves and root system.

The unique status of neem as a ubiquitous, indeed profligately abundant tree in India, coupled with its remarkable potent high activities against a truly broad spectrum of insect pests have made neem preparations technically and economically viable to the extent that to-day several commercial neem products exist in global markets. Although more than one hundred neem formulated products are available in India, there is a wide variability in concentration of registered formulations and their price. Eg. Achook, Econeem Plastic, Soluneem, Vijayneem, Neemarie and Neemark. The deterrents for their use is field instability due to rapid inactivation by sunlight, temperature etc., lower / moderate efficacy, low persistence, phytotoxicity after repeated applications, lack of quick knock down effect and reliable formulations.

There is an urgent need to characterize best ecotypes of neem tree for different environmental conditions. Intensive breeding and selection work needs to the undertaken for economic production of various high quality raw materials required for insecticide production. The development of new techniques for harvesting, depulping, drying and storage of neem seeds is a necessary pre-requisite for large scale production of neem based pesticides.

However, recent studies have revealed that neem based pesticides when alternated with conventional insecticides are as effective as the later when applied alone, against rice stem borer, leaf folder, cotton white fly and boll worm (Dhaliwal *et al*., 1996). The incorporation of neem based insecticides in the spray schedule can lead to a reduction of 40-50 per cent load on synthetic pesticides besides better control of boll worms in the cotton agro - ecosystem. Enriched neem extractives have been shown to have varied high activities against several different insect pests. Combinations of plant derivatives can enhance bio activity, thus paving the way for novel technologies using plant products and their combinations for added advantage and ecocompatibility as alternative to chemical pest control. Though research have been done on the biological activity of a number of other pesticidal plants like *Melia dubia*, *M. azeadarach*, L., *Vitex negundo* L., *Ocimum spp*, *Annona squamosa* L., *Datura* sp., *Pongamia pinnata* L., *Madhuca latifolia*, *Acorus calamus* L., *Allium sativum* L., *Capsicum* spp, *Calotropis spp*, Ginger and *Ageratum conizoides*, *Justicia adhatoda* L. they are yet to take off on a commercial scale (Saxena *et al*., 1977).

3. Mode of Action of Plant Products on Insects

Many plant products are moderately toxic to insects. They modify behaviour, exhibit antifeedant, ovpositional deterrent and growth disruptive effect. Mostly they act as contact poison, affect the structure and permeability of the cell membrane leading to leakage of cell contents from the damaged cell and the insect quickly dies. Antifeedant effect resulted in starvation death and growth disruption by prolonged developmental period, shortened adult longevity and varieties of morphological deformities.

4. Plant Insecticides & Insect Resistance

Continuous use of any insecticide, synthetic or natural may lead to the development of resistance in insects. However, Vollinger (1987) demonstrated that *Plutella xylostella* L. treated with a neem seed extract showed no sign of resistance in feeding and fecundity tests upto 35 generation when compared with deltamethrin treated lines which developed resistance. The reason is the diversity of bioactive ingredients delay development of resistance in insects.

5. Plant Insecticides and Natural Enemies

Neem products are safe to spiders, adults of numerous beneficial insect species and eggs of many predators such as Coccinellids. Nymphal or larval instars are more or less susceptible especially under laboratory conditions. No or only slight side effects are as a rule, observed under semi-field or field conditions. Due to their relative selectivity (stage selectivity) neem products can be recommended for organic pest control, as it is unlikely that they will cause severe disturbances in ecosystems. The use of neem based pesticides can be a substantial contribution towards preservation of biodiversity in ecosystems in spite of the fact that they are completely safe to all stages of beneficial organisms (Schmutterer, 1997). NSKE is safe to parasitoid, *Tetrastichus* sp, predator, *Orius* sp. Neem formulations were relatively safe to *Trichogramma australicum* giv., *Bracon hebetor* say, and *Tetrastichus isracli* (Mani & Kurian), Predator *Menochilus sexmaculatus* (Fab.) (Dhaliwal and Arora, 2001).

6. Plant Insecticides and Agroforestry

It is estimated that India has approximately 18 million neem trees with an estimated 0.7 million rabi of fruits (Narwal *et al.*, 1997) but the current production that has reached the market is only 6000 tonnes of seeds. The resource potential has been under utilized due to a change in pest control tactics and decreasing awareness about their use as pesticides. An estimated wastelands of 187.7 mllion ha in India can be reclaimed by cultivation of these trees and we can supply neem botanical pesticides to the world.

7. Limitations

Standardized bioassay procedures need to be developed for efficient screening of plants for different types of insecticidal activity. Development of suitable guidelines for registration of botanical pesticides. The mortality obtained is moderate in many cases and need to be enhanced. The pesticidal trees like neem have to be planted early in wastelands to use their products for pest management (Dhaliwal and Arora, 2001).

8. Conclusion

With the well known ill effects of chemical insecticide usage in Agriculture, Plant based insecticides offer immense scope for the pesticides of tomorrow. Again, combination of plant derivatives can enhance bioactivity (neem, sweet flag and pungam against Rice BPH.

References

Dhaliwal, G.S and R. Arora. 2001. *Integrated Pest Management, Concepts and Approaches. Kalyani Publishers,* New Delhi. PP. 427.

Dhaliwal, G.S., J. Singh and V.K. Dilawari, 1996. Potential of neem in insect pest management in rice. In: R.P. Singh, M.S. Chari, A.K. Raheja and W. Krau (eds). Neem and Environment. Vol 1. Oxford & IBH Publishing Co., Pvt. Ltd. New Delhi, PP. 425-431.

Narwal, S.S., P. Tauro and S.S. P. Tauro and S.S. Bisla, 1997. Neem in Sustainable Agriculture Scientific publishers, Jodhpur. PP. 266.

Saxena, B.P., O. Koul, K. Tikker and C.K. Atal 1977. A new insect chemosterilant isolated from *Acorus calamus* L. Nature. 270: 512-513.

Schmutterer, H. 1997. In CRC Handbook of Natural pesticides Vol III Insect Growth Regulators, part B. (Eds.) Morga, E.D and N.B Mandava. CRC, Boca Raton, Florida, PP. 119-170.

Singh, R.P., S.Singh and S. Wahab. 1999. Biodiversity and importance of botanical insecticides. In: G.S. Dhaliwal, R. Arora and A.K. Dhawan (eds.) Emerging Trends in sustainable Agriculture. Common wealth Publishers, New Delhi, India PP. 1-9.

Vollinger, M. 1987. The possible development of resistance against neem seed kernel extract and deltamethrin in *Plutella xylostella*. In H. Schmutterer and K.R.S. Asher (eds.) Natural pesticides from the neem tree (*Azadirachta indica* A. Juss) and other tropical plants. Proc., 3rd Intl. Neem conference, Nairobi, Kenya. PP. 543-554.

❑❑❑

SECTION - V

PHEROMONES AND TRAPS

22

Use of Pheromones for Managing Pests of Field Crops

NANDAGOPAL, V.
Principal Scientist (Entomology) Central Rice Research Institute (ICAR)
Cuttack-753 006, Orissa

1. Introduction

Chemicals used for communication between the organisms are known as semio-chemicals or info-chemicals. These chemicals are also called "assembling scents". In Greek *Pherein* means to carry and hormone (to excite, stimulate) as described by Karlson and Butenandt (1959). Further Kirschenblatt (1957) proposed the term "*telegones*". Micklem (1959) termed it as "*pherormone*" instead of pheromone. Karlson and Luscher (1959) have given a more scientific clarification for the use of the term *pheromone*.

Pheromone has been in use in India over more than 30 years as a monitoring tool under All India Coordinating programme for some of the crops likes cotton, groundnut and pigeon pea. The monitoring technology is well knitted in different centers in these crops. The monitoring helps in developing forewarning systems and plan management strategies. Unfortunately, as such efforts have not been made even at research station; the farmers are not convinced about practical utility of pheromone technology. Some farmers use pheromone traps especially in cotton and pigeon pea eco system, as a mass-trapping tool. Surprisingly, in a recent study at the NRCG; a maximum of 1470 males/day/trap in case of *Spodoptera litura* was recorded showing its potentiality for mass trapping.

2. Types of Pheromones Reported

There are, at present ten different types of pheromones identified world wide. They are

- Territorial pheromones
- Trail pheromones
- Alarm
- Aggregation
- Sex pheromones
  a) Attract and kill
  b) Lure and infect -Auto-infection
  c) Monitoring
  d) Mass trapping
  e) Mating disruption
- Epideictic (spacing) pheromones
  a. Oviposition deterrent pheromones
  b. Mating deterrent pheromones
- Nasonov pheromone
- Royal pheromones
- Calming (appeasement) pheromones
- Primer pheromones

In India, monitoring of the key insect pests such as *H. armigera, S. litura, P. gossypiella,* were undertaken as early as 1985. Intensive work has been done on these insects spreading over Karnataka, Haryana, Madhya Pradesh, Tamilnadu, Andhra Pradesh, Maharashtra, Himachal Pradesh, Punjab and Gujarat. Pheromone as a mass trapping tool have also been utilized in a few insect such as *Pthorimoaea opercullela, S. litura, Chilo sacchariphagus indicus, H. armigera, A. modicella, Lymantria rabica* Wlk (Porthetria dispar), Tortricid *Cydia pomonella* (L.), Codling moth, *Cydia pomonella,* and *Pectinophora gossypiella* (Saund.). The communication disruption as a tool has been tried in *S. litura, A. modicella, Peripleneta rabica* and *Chilo auricilius* in India with limited success. There are attempt made to design the trap for efficient trapping of the target insect, some of them were successful in the case of cotton, sugarcane and groundnut ecosystem. Interestingly, lot of work has been done on the basic aspects of pheromone since 1975.

3. Synthesis of Pheromone

The chemistry part of the pheromones is in its developing stage in India as sophisticated instruments are required for taking up of Isolation, Identification and synthesis of pheromones. Very few labs such as Pheromone groups of IICT, Hyderabad, Organic chemistry group of SPIC, Organic chemistry group of BARC have such facilities in India. Some works on the sex pheromone of the crop pests in India has been taken up in collaboration with foreign scientific organizations such as NRI, London; TNO, Delft, Netherlands, NL and CID, CSIC, Barcelona, Spain etc.

4. Monitoring

Where Monitoring can be Applied?

Timely and efficient monitoring should be the foundation of sound IPM programs, without which no proper pest management decision can be made. A few of the examples of application of pheromones employed for monitoring purposes are as follows:

- Early detection of pest
- Mapping pest distribution
- Quarantine inspection
- Pest monitoring for action thresholds / decision support
- Mapping insecticide resistance frequency
- Monitoring parasites / predators
- Estimation of population dynamics
- Estimation of population densities
- Timing of management procedures based on the threshold
- Efficacy of management procedures based on the threshold

Pest monitoring using sex pheromones facilitates early detection of the pest occurrence, through moth population dynamics during crop season and this information helps in making decisions on the choice of pressing control tactics in other words precise timing of insecticide or bio-agents applications.

Several parameters enable of efficient pheromone based monitoring for insect pests of rice crop. Three traps per acre as trap density for monitoring of rice yellow stem borer were optimized. Various parameters of an efficient pest monitoring system for rice pests are presented.

4.1 Monitoring System in Cotton

Several studies have been conducted on cotton pests to establish a quantitative relationship between pheromone trap catches of particular pest and the damage caused with an aim to develop relationship would help to forewarn. Utilization of pheromones of various cotton pests as follows (Table 1).

Table 1: Utilization of pheromones of various cotton pests as follows

Bollworm trap	ht	Predictability of larval damage	catch/night/trap
Bollworm		for 3 nights	Expect egg laying in the next 5-8 days
Pink boll worm /night	0.03m	8-10males/trap for 4 nights	Expect egg laying in in the next 10-15 days
Spotted bollworm	0.03m	1 males/night for 3	Expect egg laying in in the nights next 3-5 days
Spiny bollworm	0.03m	1 males/night for 3 nights	Expect egg laying in in the next 3-5 days

4.2 Monitoring System in Groundnut

The presence of sex pheromones of leaf miner has been identified for the first time by the National Research Centre for Groundnut (NRCG) in collaboration with NRI, U.K. and field evaluation of the components of the sex pheromones have been also done by the NRCG. Experiments were also conducted by NRCG for standardizing the number of traps per unit area and use of the sex pheromone as one of the components of Integrated Pest Management (IPM) in groundnut where leaf miner damage is extensive under Saurashtra conditions. Nandagopal and Soni have evaluated a number of traps for the groundnut leaf miner (*Aproaerema modicella* Deventer. Hall *et al.* have identified the total components of the pheromone compounds of leaf miner.

At Junagadh, a square glue trap developed at NRCG was used, while in other centres water trap designed by ICRISAT was used. Different types of traps were designed and tried in the trapping of males of *A. modicella*. The results indicated that a square glue trap designed with 22 cm side and with 4 entry holes was the most efficient in trapping the males.

Nandagopal *et al.* reported on the population dynamics of *Spodoptera litura* (F.) in relation to weather parameters in groundnut in India using pheromone trap catch. Nandagopal have developed an efficient pheromone trap for field catch of *Spodoptera litura* Fabricius in groundnut and castor ecosystem. Based on the pheromone trap catches the prediction of *Spodoptera litura* (F.) males in relation to key weather factors in groundnut have been reported. Holistic IPM wherein the use of pheromone has been explained.

4.3 Monitoring System in Red Palm Weevil *Rhynchophorus ferrugineus* Oliv.

Attraction of red palm weevil Rhynchophorus ferrugineus to ferrugineol based pheromone lures in coconut.

4.4 Monitoring of Coffee White Stem Borer: X*ylotrechus quadripes* (Coleoptera: Cerambycidae)

Biological methods of control proved to be a failure so far. Therefore, an alternate eco-friendly technique like use of sex pheromone has been developed after taking a clue that these beetles communicate each other through chemical medium. Studies showed that male borers attract the females by secreting sex pheromone. The sex pheromone components from the males have been successfully isolated, chemically characterized and synthesized in the laboratory. The components were identified to be (*S*)-2- rabica-3-decanone, I-3-hydroxy-2-decanone and 2,3-dihydroxy octane.

Mass Trapping

This is a direct method of managing the insects. Based on the number of insects trapped, a decision can be taken to continue the recommended number of traps/unit area for mass trapping or remove all the traps and use only for monitoring. The following are the insects for which mass trapping was tried (Table 2).

Table 2: Various insects against which mass trapping was tried

Species	Location
Pthorimoaea opercullela	Karnataka and Tamil Nadu
	Karnataka
	New Delhi
	Uttar Pradesh
	Punjab
Chilo auricilius	Uttar Pradesh, Hariyana
S. litura	Uttar Pradesh
	Andra Pradesh
	Tamil Nadu
	Gujarat
	Gujarat
H. armigera	Andra Pradesh
	Gujarat
	Haryana
	Gujarat
	Tamil Nadu
	Tamil Nadu
	Gujarat
	Gujarat
Aproaerema modicella	Gujarat
	Andra Pradesh
Lymantria mericana (=*Porthetria dispar*)	Jammu Kashmir
Cydia pomonella	Jammu Kashmir
	Ladakh
Pectinophora gossypiella	Tamil Nadu
	Gujarat
Periplaneta mericana	Karnataka
Orcytes rhinoceros	Andra Pradesh
Rhynchophorus ferrugineus	Goa
	Karnataka
Scirpophaga incertulas	Andra Pradesh
Xylotrechus quadripes	Karnataka

Chilo sacchariphagus indicus	Karnataka
Scirpophaga rabicapid	Karnataka
Chilo infuscatellus	Karnataka
Batocera dorsalis	Gujarat Gujarat
Odoiporus longicollis	Tamil Nadu
Earias spp.	Tamil Nadu Gujarat
Leucinodes orbonalis	Gujarat
Cylas formicarius	Kerala
Plutella xylostella	Karnataka

Mating Disruption

Male confusion is thought to be the result of ambient pheromone concentrations sufficient to hide the trails of the calling female to the large doses from diffuse sources such as microcapsules or larger doses of pheromone at the point source dispensers such as tie-on polyethylene rope.

Some possibilities of the mechanism of the confusion approach as put forth by Pedigo include:

- Camouflage or covering up the natural pheromone scent of females
- Misdirection of males to scents from multiple point sources of synthetic pheromone, and
- Adaptation/habituation by desensitizing male antennal receptors through constant pheromone-like exposure

5. Trap Design

Limited information is available on the development of traps for various insects in Indian agriculture. There has been efficient trap design for groundnut leaf miner in groundnut ecosystem, for *H armigera.* For sugarcane internode borer, *Chilo sachariphagus indicus* Kapur for *S. litura.* The efficiency of these traps were between 60 to 90 %. In the case of *Pectinophora gossypiella* (Saunders) in cotton ecosystem an interesting observation was reported. The trap and the lures used for

H. armigera attracted the *Pectinophora* males. Reed *et al.* and Satpute *et al.* reported a efficient trap for *P. gossypiella.*

5.1 Various Types of Traps Available are as Follows

- Jackson traps are preferred for monitoring the fruit flies
- Ramp traps are used for ground-crawling beetles and weevils.
- Multiple-funnel traps are the best choice for monitoring and mass trapping coffee berry borer (*Hypothenemus hampei*).
- Wing traps are used for short-term monitoring of some moth species.
- Diamond traps are used for monitoring moths in storage areas.
- Unitraps are preferred over other designs for long-term monitoring and mass trapping of moths in the field and in storage areas.
- McPhail traps are preferred for monitoring and mass trapping fruit flies.
- Vane Traps are excellent traps for mass trapping scarabs.
- Probe traps are excellent for monitoring grain beetles in bulk storage facilities.
- Boll weevil traps are specifically designed for monitoring and mass trapping cotton boll weevil (*Anthonomus grandis*).
- Grain Beetle Surface Traps are referred for monitoring the populations of beetles and weevils in stored products.
- Khapra Beetle Wall Traps traps are specifically designed for monitoring Khapra beetle, *Trogoderma granarium* in stored products.

Source : http ://www.agbio-inc.com/

5.2 Trap Design Developed

Information available on the development of traps for various insects in Indian agriculture is limited. Efficient traps were designed for *H armigera* groundnut leaf miner Sugarcane internode borer, *Chilo sachariphagus indicus* and *Spodoptera litura.* The efficiency of these traps were between 60 to 90%.

There are many kinds of traps available for various important insect and mite pests. These include Sleeve trap, Water trap made of Galvonized Iron trays or Plastic Watta -Trap, Sticky trap such as Delta trap, Diomond trap, Panel trap, Unit trap, Rhegalatis trap. The plate 3. is the stika trap named David's trap has been developed for groundnut leaf miner in early 1990s.

There are particular trap for each of the species of insects belonging to a family/order to trap a specific pest. For example water traps are mostly used for Pyralid moths of the order Lepidoptera. When it was modified using a plastic, the trap is used for catching males of brinjal fruit and shoot borer and Groundnut leaf miner a Gelechiidae of order Lepidopera.

Reference

Nandaghopal, V., J. Anand Prakash, Rao, J.S. Yadav and A.R. Prasad. 2007. Pheromones: Principles and Practices, Central Rice Research Institute, Cuttcak, Indian Institute of Chemical Technology, Hyderabad, Applied Zoologist Research Association, CRRI, Cuttack. pp. 350.

23

Traps for Management of Stored Product Insects

MOHAN, S. AND P. PRETHEEP KUMAR
Department of Agricultural Entomology,
Professor of Entomology, Dept. of Cotton, Tamil Nadu Agricultural University
Coimbatore – 641 003, Tamil Nadu

1. Introduction

Insect pests have been a threat to food and seed since man has started growing crops. Almost all stored-products and seeds are liable to insect injury. The protection of stored grains from insect pests is of considerable importance to their severe infestation and damage in a short period. Often the presence of insects in store houses are felt only when they are hovering and flying around, by which time enormous loss and population build up of insects might have occurred. It is well proved that no granaries can be filled with grains without insects as the harvested produce contain egg (or) larvae (or) pupae in them because of field carryover infestation which cannot be avoided in developing countries like India. Hence, development of simple technologies for timely detection of insects in the stored produce to take up timely control measures is very important (Mohan and Gopalan, 1992).

2. Trapping Devices

Insect traps are excellent insect detection devices in food grains and are more effective in the detection of stored grain insects namely

Rhyzopertha dominica (F.), *Sitophilus oryzae* (L.) and *Tribolium castaneum* (Herbst). They are as follows:

- TNAU Insect probe trap
- TNAU Pitfall trap
- TNAU Two-in-one trap for pulse beetle
- Indicator Device
- Automatic insect removal bin
- UV - Light trap for warehouse
- Stored grain insect pest management kit

All these devices can be used for both monitoring and mass trapping of stored grain insects. It is important to note that the presence of even a single live insect in food grain cannot be tolerated as they build up and cause enormous loss in storage due to their high reproductive rate. Farmers immediately after harvest dry their produce below 10% MC (moisture content) and store them in bags and bins, etc. These harvested produce contain insect life stages, i.e. egg, larvae (or) pupae (which are inside the grain) which are carried over from field. It is well proved by researchers that this field carry over damage is <1% but if it is not controlled total loss can be expected within 2 - 3 months. It is well known that these field carryover infestation adults generally emerge 15 - 20 days after storage and start mating, laying eggs and thereby multiply into large numbers. The traditional practice followed by farmers is "SUN DRYING". They do it generally after a month of storage of food grains by which time the eggs of the insects hatch and enter the grains and get concealed in many cases. Sun drying can kill only eggs and not the larvae (or) pupae which are concealed inside the grain. Hence, there is a need to timely detect the emergence of field carry over infestation during storage and to take up timely operations like sun drying which will help to kill the eggs and thereby prevent multiplication. Trapping devices could be used for detection as well as control of stored product insects in bag or bin storage (Mohan, 1994, 1996, 1997).

2.1 TNAU Insect Probe Trap

The basic components of a TNAU probe trap consists of three important parts: A main tube, insect trapping tube and a detachable cone at the bottom. Equispaced perforations of 2 mm diameter are made in the main tube.

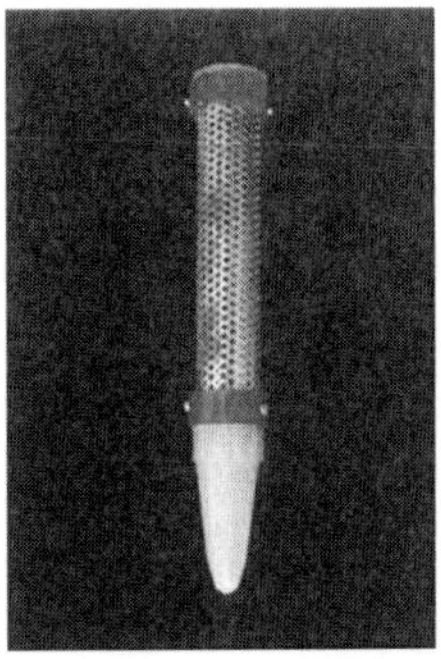

Wandering behaviour and migration of insects towards aeration is exploited in this technology. The insect trap has to be kept in the grains like rice, wheat etc., vertically with the wite plastic cone downside as shown the figure. The top red cap must be with the level of the grain. Insects will move towards air in the main tube and enter through the hole. Once the insect enters the hole it falls down into the detachable white cone at the bottom. Then there is no way to escape and the insects are trapped forever. The white detachable cone can be unscrewed once in a week and the insects can be destroyed. Probe traps should be placed at top 6 inches of the grain, where the insect activity is seen during early period of storage. They can remove > 80% of the insects within 10 - 20 days.

Merits

- No chemicals.
- No side effects.
- No maintenance cost.
- Easy handling.

2.2 TNAU Pit Fall Trap

Pitfall traps are used for capturing insects active on grain surface and in other layers of grain. (Monitoring and mass trapping tool).

Standard Model

It has 2 parts, namely, a perforated lid (2 mm (or) 3 mm) and a cone shaped bottom portion. Application of a special coating with sticky material on the inner side of cone to hold trapped insects is necessary in this model and this procedure is tedious.

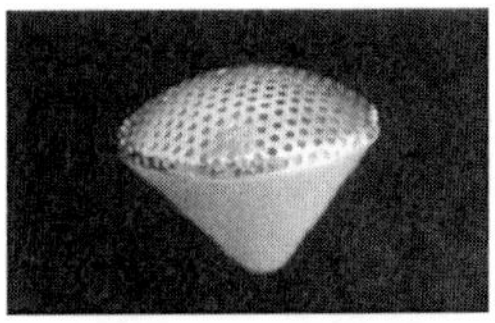

Standard Model

TNAU Model

It has a perforated lid, cone shaped bottom which tapers into a funnel shaped trapping tube and no sticky material need to be applied in this model.

TNAU Model

2.3 TNAU Two-in-One Model Trap

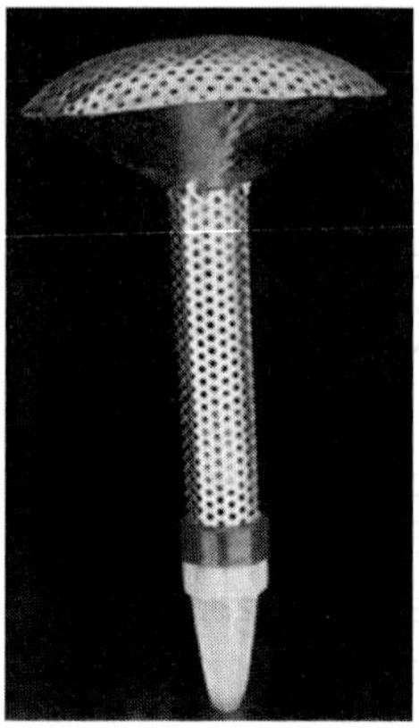

The probe trap containing the components namely the perforated tube, pitfall mechanism, a collection tube and the cone shaped pitfall trap with a perforated lid and the bottom tapering cone were combined as a single unit. Combination of probe and pitfall increase the trapping efficiency of insects. Best suited for pulse beetles as they are seen only on grain surface wandering here and there. It does not require tedious procedures like coating the inner surface of pitfall cone with sticky materials before trapping to hold pulse beetles. Beetles are captured alive in this trap, which may facilitate release of pheromone and there by attract more insects.

2.4 Indicator Device

It consists of a cone shaped perforated cup (3mm perforation) with a lid at the top. The cup is fixed at the bottom with a container and circular dish, which are to be smeared with sticky material like rabica. Farmers, before storing their pulses, should take 200 g of pulses to be stored and put them in the cup. When the field carried over beetles start emerging, due to their wandering behaviour, they enter the perforations and get slipped off and fall into the trapping portions. As they stick on to the sticky materials, farmers can easily locate the beetles and can take out the bulk-stored pulses for sun drying. The device with 2mm perforations can be used for cereals. This will help in eliminating the initial population, which acts as the major source for further build up. Thus, timely detection will help the farmers to preserve their valuable pulses during storage.

2.5 TNAU Automatic Insect Removal Bin (Mohan, 1995)

TNAU automatic insect removal bin has 4 major parts, namely, outer container, inner perforated container, collection vessel and the lid. The model exploits wandering behaviour of stored product insects as well as the movement of these insects towards well aerated regions. The grains are held in the specially designed inner perforated container. The space between inner and outer container provides good aeration for the insects. Insects, while wandering, enter the perforation to reach the aerated part and while doing so, get slipped off and fall into the collection vessel through a pitfall mechanism provided in the collection vessel. In order to quickly collect the insects, as and when they emerge from grains, perforated (2 mm) rods are fixed in the inner container.

The container will be useful for storing rice, wheat, broken pulses, coriander, etc. The insects such as rice weevil, lesser grain borer, red flour beetle, saw toothed beetle, which are commonly found attacking stored grains can be removed automatically by storing grains in this container. Within a very short period of 10 days a majority of the insects (more than 90 per cent) can be removed from the grains. The containers are available in 2 kg, 5 kg, 25 kg, 100 kg and 500 kg capacities.

Automatic insect removal bin can remove substantial number of emerging adults and sometimes free living larvae also.

2.6. UV Light Trap for Grain Storage Godowns

The UV light trap mainly consists of an ultra-violet source (4 W germicidal lamp). The lamp produces ultra-violet rays of peak emission around 250 nm. The light is fitted at the centre of a funnel of 310 mm diameter at the top and 35 mm diameter at the bottom. The bottom end of the funnel is attached with a transparent plastic container for collecting the trapped insects. To hang the unit at desired points, three hooks have been provided at the periphery of the funnel. The unit is also provided with a tripod stand.

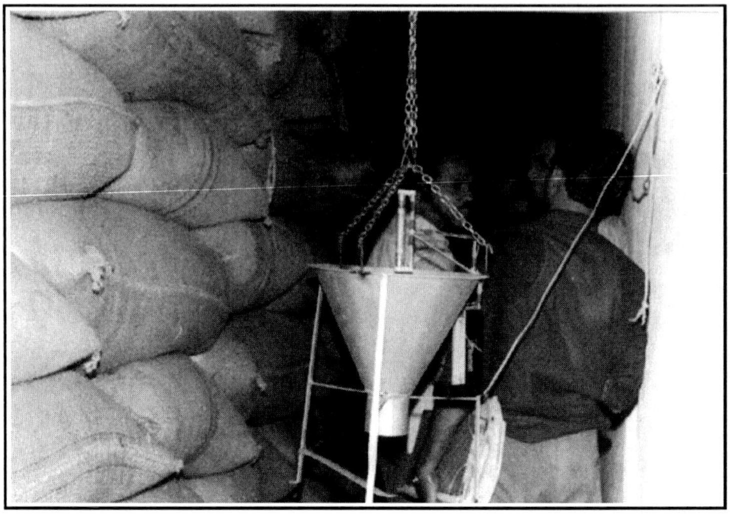

The UV light trap can be placed in food grain storage godowns at 1.5 m above ground level, preferably in places around warehouse corners, as it has been observed that the insect tends to move towards these places during the evening hours. The trap can be operated during the night hours. The light trap attracts stored product insects of paddy like lesser grain borer, Rhyzopertha dominica, red flour beetle, Tribolium castaneum and saw toothed beetle, Oryzaephilus surnamensis in large numbers. Psocids which are of great nuisance in godowns are also attracted in large numbers. Normally 2 numbers of UV light trap per 60 x 20 m (L x B) godown with 5 m height is suggested.

The trap is ideal for use in godowns meant for long term storage of grains, whenever infested stocks arrive in godowns and during post fumigation periods to trap the resistant strains and left over insects to prevent build up of the pest populations. In godowns of frequent transactions the trap can be used for monitoring. It has been found that two traps kept at the corners of the warehouse (60m x 20m x 5m) can catch around 200 insects/day even from a godown where normal sampling did not show any insect presence, thus indicating its effectiveness as a monitoring and mass trapping device.

2.7 TNAU Stored Grain Insect Pest Management Kit

Food grains are stored for varying periods to ensure proper and balanced public distribution throughout the year. Among the biotic and abiotic factors which affects grains / seeds in storage, insect plays a major role in the deterioration of grains/seeds causing both qualitative and quantitative losses. Hence, timely detection of the stored grain insects will help to prevent heavy losses.

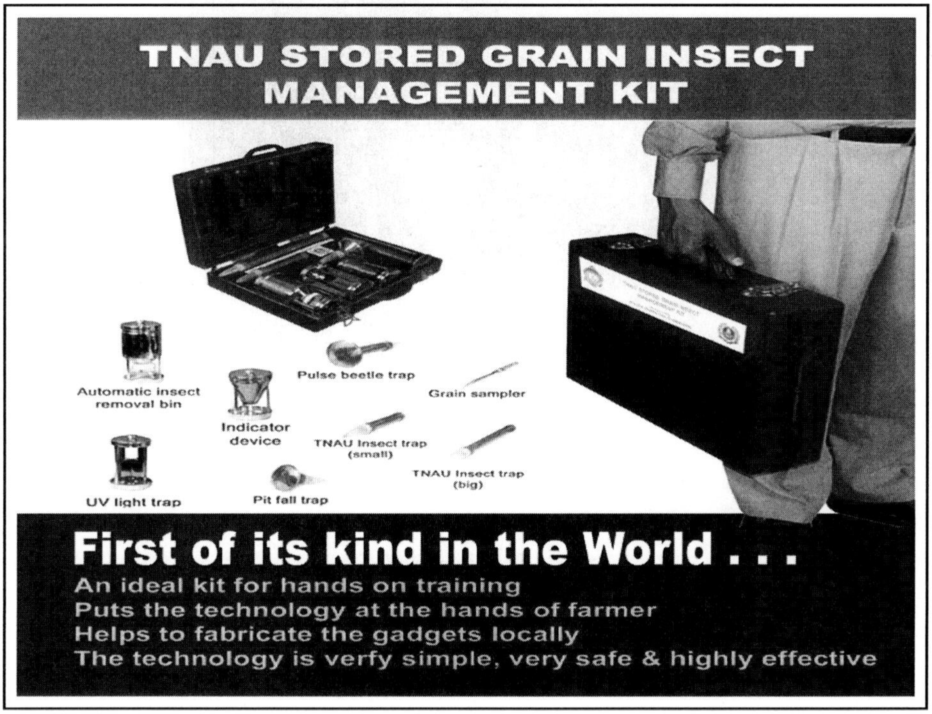

Tamil Nadu Agricultural University (TNAU) is one of the pioneering institutes in India in the development of detection devices for stored grain insects. These devices exploit the wandering behaviour of the insects and help in timely detection of insects in stored produce leading to timely control. These include TNAU probe trap, TNAU pit fall trap, two in one model trap, indicator device, Automatic insect removal bin and UV-light trap technology. These devices have been widely used in many places and have received State and National recognitions.

Hence, the Department of Agricultural Entomology, Centre for Plant Protection Studies, TNAU, Coimbatore has developed a "KIT" named as TNAU-Stored Grain Insect Pest Management Kit containing prototypes of all the devices along with a CD-Rom about the devices and how to use them. This kit will be of great use in popularization of the technologies across the country. The kit will be an ideal "hands-on training" tool for Education, Extension centers (KVK, Plant clinic, save grain centers) and also for private ware housing. This TNAU kit is the first of its kind in the world.

References

Mohan, S. and M. Gopalan. 1992. Stored grain insect trap. *Rice India*, 2(8): 15-16.

Mohan, S. 1994. A storage container model for automatic removal of stored product insects from grains. *Bull. Grain Tech.*, 32(2): 178-181.

Mohan, S. 1995. A storage container for automatic removal of insects from grains. *Indian Food Ind.*, 14(3): 45.

Mohan, S. 1996. A new storage bin. *TNAU Newsletter*, 26(1): 2-3.

Mohan, S. 1997. Laboratory studies of a new storage bin to remove *Sitophilus oryzae* L. and *Rhyzopertha dominica* F. from stored paddy. *Pestology*, 21(12): 30-35.

❑❑❑

24

Use of Lures and Traps as Ecofriendly Techniques for Organic Pest Control : Case Study of Fruit Flies

SITHANANTHAM. S.
Director, R & D, Sun Agro Biosystems Ltd., Porur, Chennai – 600 116

1. Introduction

Insect pest management with emphasis on organic practices calls for eco-friendly and timely control in our agricultural crops. Although the use of insect attractants as lures and traps with or without attractments is an age - old practice recent research and development has led to a wider array of such options. Developing upon the local crop requirements, these could be deployed for monitoring or mass trapping of the different target pests.

2. Feeding Attractants / Lures

The more common attractants for insect pests are based on food attraction ie. sugar -source (eg. jaggery, molasses) and fermentation products (eg. Protein hydorlysate, banana fruit, yeast linked sources like vinegar) These are used as additives to poison baits (eg. for cutworms) or as lures in monitoring traps (eg. Fruitflies). Baits include base material like grains/ flour, plus attractan (like jaggery) and poisons (Synthetic insecticides) in suitable proportion.

3. Sex Attractants/Pheromonse /Lures

In the last 2-3 decades, the use of sex attractants as lures for attracting the opposite sex adults of the pest is gaining importance. Personally, the commercially produced synthetic attractants are mostly for males of Lepidoptera besides for adults of coleopteran and Dipteran groups. The target pests for which the lures are presently available in India are listed below.

S.No	Pest	Scientific Name	Crops
1	American Bollworm	*Helicoverpa armigera*	Cotton, Pulses, Vegetables
2	Cotton Leaf Worm	*Spodoptera littoralis*	Cotton
3	Yellowstem borer	*Tryporyza incertulas*	Rice
4	Leaf Roller	*Cnaphalocrocis medinalis*	Rice
5	Striped Bollwtrm	*Earias insulana*	Cotton, Okra
6	Pink Bollworm	*Pectinophora gossypiella*	Cotton
7	Early shoot Borer	*Chilo infucatellus*	Sugarcane
8	Internode borer	*Chilo sacchariphagus*	
9	Topshoot borer	*Tryporyza excerptalis*	
10	Shoot and Fruit borer	*Leucinodes orbonalis*	Brinjal
11	Diamond backmoth	*Plutella xylostella*	Cabbage, Cauliflower
12	Stalk borer	*Chilo auricilius*	Sugarcane
13	Rhinoceros beetle	*Oryctes rhinoceros*	Coconut, Oilpalm
14	Red palm weevil	*Rhynchophorus ferrugineus*	Coconut

4. Parapheromones

These are a small groups of compounds, designated so because they are not produced by the target species but elicit attraction in them, other than as food attractants. They are like aggregation pheromones in their action, while predominantly attractive to adult males among the fruit flies Cfam: Tephretidae: order: Dipera). Examples of this category are listed below Methyl eugeol: For Bactrocera dorsalis B. Correcta, Bzonata (in many fruit crops) Cue lure: For B. Cucurbital (incucurbit vegetables) Med Lure and Trimed lure: For mediterraneam fruit flies.

5. Light Traps

These are of different designs depending on power/light source and whether they are for monitoring or mass trapping. Burning of waste stalks in fields or keeping burning torches could also serve as light for short spells in fields. Scarabacid Beetles (adults of white grubs - different genera) and moths (adults of redhairy caterpillar are attracted by light sources and used for mass trapping during their peak emergence periods as a campaign action. The use of incandescent bulbs as in Robinsons light trap is often used for surveys and season long monitoring of many important pests (mainly beetles/ moths).

6. Suction Traps

These require electricity and used for continuous or periodical monitoring of air borne insects. Small sized and wringed stages of insects like aphids and thrips have been successfully sampled for seasonal pattern monitoring in bench mark locations. They consist of a suction fan, funnel and collecting jar and kept stationary.

7. Water Traps

These are used in field crops in which open space will be available to install and service them. They consist of a coloured trap/ pan kept at or near ground level and filled up to the fourth of the depth with water. Soap solution or mineral oil is added to the top layer to minimise evaporation of the water. Yellow colored trays/ pans are used for attracting thrips, aphids, hoppers and small beetles among others. Water traps are also useful for trapping the adult moths attracted to pheromone lures of sugarcane borers.

8. Stickly Board /Delta Traps

These consist of a durable gum coated card board surfaces to intercept and retain small flying insects in crop systems. For example yellow sticky boards (about 10X16 inches) when suspended in crops like cotton, grapes, mangoes can help trapping aphids, thrips, hoppers etc. Delta traps are triangular in cross section and used for capturing small sized moths (like Plutella) when attracted to sex pheromone lures.

9. Mc Phail Traps

These are mainly used for capturing adult fruitflies. They are cylindrical or pear shaped, with a circular window at the bottom. The parapheromone lures are fixed or suspended inside these countaiensr at the top and their plume attracts the adults. After the adults enter from below they are unable to seek exit, since they tend to go up towards the lune (and the light) above.

10. Bucket Traps

Big plastic buckets (about 10-15 lit.capacity are used in the four rectangular windows (about 2X4 inches) cut at the opposite sides, about 2-3 inches below the top rim. The attractant (Pheromone) plus food bait (banan other fruits are kept at the bottom of the bucket. These are presently popular for the rhinocrros beetle and red palm weevil. They are tied at about 3-4 feet above the ground on the palm trunk. The outer surface of the buckets can be glued with gunny jute) cloth to provide grip to the landing beetles so as to minimize escapes.

11. Recent R&D on Fruitfly Traps/Lures

Improvements have been made in choice of useful combinations of color and shape of the trap containers for mango/guava/orange crop attacking fruit flies. It is also found that the attractant (methyl eugenol) can be dispensed in a wick system instead of in soaked ply woods to enhance the catches as well as to extend the duration of coverage. This will minimize / avoid the need for refill/ replace the lure within the fruiting season (about 8-10 weeks). At sun Agro Biotech a Research Centre (Sabro Chennai) further research is going onto determine and possibly enhance the cashing efficiency for each of the major fruitfly species (*B. dorsalis, B. corecta , B. eoreta*) and to minimize escapes in the traps. These are important initiatives towards establishing "fruit fly free zones" which are a requirement to qualify for mango export in India.

❒❒❒

25

Traps in Crop Pests Management

KATHIRVELU, C.
Senior Lecturer, Dept. of Entomology, Faculty of Agriculture,
Annamalai University, Annamalainagar-608002, Tamilnadu

A trap is a device that impedes or stops the progress of an organism. These traps provide valuable information on the insect population fluctuations and their peak period of emergence as well as their activities. Traps are also useful in mass trapping and killing of insects. The various types of traps have been developed based on factors like physical (light and sticky traps), chemical (pheromone and bait traps) and mechanical (suction, pitfall and emergence traps). Currently they are employed by the entomologists in the integrated pest management of many agricultural insect pests.

1. Light Trap

The first known record of attraction of some adult insects to artificial light was observed and recorded by the Greek poet Aeschylus (525-456 B.C.) in his Fragments 288 (C.450 B.C). In earlier days bon fires were used for attracting and killing insects. This was followed by kerosene burning lamps and lanterns as source of attractants. In a number of cases these lamps or lanterns were used inside a room or other sheltered location where they attracted moths were trapped and caught by hand.

1.1 Sources of Light

Numerous lamps were used in traps to determine the photo responses of various economically important insects during the past 25 years. These lamps provide a wide range of radiation output in terms of quality (wavelength) and quantity (power). These lamps can be categorized into three major groups *viz.*, inside frosted incandescent lamp, mercury vapour lamp and black light.

1.1.1 Incandescent Lamps

They produce radiation by heating a tungsten filament. The spectrum of lamp include a small amount of ultraviolet, considerably visible light especially rich in yellow and a peak of radiation in the infrared region which includes about three fourth of the total lamp output.

1.1.2 Mercury Vapour Lamps

They produce primarily ultraviolet, blue and green radiation with little red. The proportion of infrared radiation is much smaller than that of incandescent lamps. A modification of this type of lamp is fluorescent lamp. This is also a mercury vapour lamp with the inner surface of the lamp coated with various phosphorus compounds which observe short wavelengths and reradiate the energy at longer wavelength. The spectral output of this modified mercury vapour lamps include the continuous spectra from the fluorescing phosphorous plus the bright light as in the non coated mercury vapour lamps.

1.1.3 Black Light

Black is the popular name for ultraviolet radiation energy with the range of wavelengths from 320 to 380nm. This is the most common light source used in many insect control and survey programme in developed countries.

All available models of light trap use one of the above lamps as a source of attraction. The various models of light trap available are grouped according to the sources of light used: incandescent light, mercury vapour light and black light trap (UV light trap).

1.2 Types of Light Traps

1.2.1 Incandescent Light Trap

1.2.1.1 Primitive Light Trap Model: Montana Light Trap

The trap consisted of galvanized iron wash tub and lantern. A galvanized iron arch was fitted across the tub, which served to deflect the moths and to hold the lantern. This was used to trap cutworms, *Agrotis ipsilon* in California during 1921. The lantern emits light rays consisting of visible spectrum and infra red (Parker *et al.,* 1921).

1.2.1.2. Simple Incandescent Light Trap Model

These traps have pans filled with water and thin film of oil. It is kept below an electric lamp fitted to a reflector. This type of trap was very simple consisting of a reflector, lamp and water pan.

1.2.1.3 A Low Intensity Incandescent Light Trap

It is for selective capture of smaller delicate insects designed by Mohan *et al.* (1993). It is a 3000ml plastic container cut into two equal parts. On the top portion equispaced perforations of 15mm diameter are made into two rows. These perforations are equally distributed on all sides and act as a screen for selective capture of delicate and smaller insects. A low incandescent light source (40 Watt) is mounted at the top portion. At the bottom of the top portion a plastic funnel with its stem cut short to one centimeter length is fixed. The top portion, funnel arrangement and the bottom collection unit are held tight in position by a stand arrangement. This trap is highly useful for monitoring rice gall midges, brown plant hopper, white backed plant hopper, leaf hoppers, *Culicoides* sp. (Biting midges) (Mohan *et al.,* 1993).

1.2.2 Mercury Light Trap Models

1.2.2.1 Robinson Trap

The trap is the basic model designed by Robinson (1952). The trap proper is basically an upper structure consisting of an inverted truncated cone containing a number of radial vanes at the centre, with a light source lying just above the plane through the top of the cone. The collection device used is a parcel box in which cotton soaked in insecticide is placed. In order to prevent the escape of insects the cone is mainly constructed of transparent material so that the point of exit may be invisible from the interior. This is currently used towards a wide range of noctuids and other nocturnal flying insects.

1.2.2.2 Muguga Light Trap

This is similar to Robinson trap. The truncated cone of clear plastic admitting light to the interior from around the rim of Robinson trap was replaced by a sheet metal cone carrying the lamp and fins. The lamp was protected from rain by an inverted two-litter glass beaker. The light source is 500W mercury bulb. Insects belong to family Pyralidae, Sphingidae and other Noctuidae were attracted using this trap (Taylor and Brown, 1972).

1.2.2.3 Rothamsted Light Trap

It has 960cm square roof over the lamp confining the illumination to a beam which diverges about 22.3° above the horizontal in all compass direction. It can be fitted with 200W tungsten lamp or 125W mercury lamp. Lepidopteran moths of family Sphingidae, Pyralidae, and other Noctuidae were attracted using this trap (Taylor and Brown, 1972).

1.2.3 Black Light Trap Models

1.2.3.1 Black Light Insect Survey Trap

The basic components of a black light trap model consists of baffle assembly with four baffles (two crossed), black light lamp (generally 15 W), funnel (60° slope) collection container, tripod and electrical components. It can be used for a wide variety of nocturnal insects for obtaining indices of their population levels (Harding *et al.*, 1966).

1.2.3.2 Electric Grid Type Trap Model

The light trap with electric grid is flat or a hollow cylinder surrounding the lamp. This flat grid type has been used where black light lamps are used as attractant. This was used for attracting *Heliothis zea* in corn field in California. The problem faced here was serious clogging of the grids when heavy flights of European corn borer is noticed in fields. The grid spacing was 3/8 inch with a voltage of 3,500 - 4,500 (Heinton, 1960).

1.2.3.3 Electrocute Grid Insect Trap

Electrocuter grid insect trap known commonly as "grid trap" is used in applications intended to suppress nuisance insects around residence, business establishments and food processing plants. Generally black light (BL) lamp is located relatively close to the charged grids so that attracted insects come into contact with the grid.

It consist series of adjacent wires insulated from each other and connected to opposite polarities of a low current high voltage transformer. The assembly of charged wire is most frequently made in the form of a flat panel or circular case with the provision for an attractant lamp. An insect flying between the wires draws an arc that passes through its body usually causing death (Mitchell *et al.*, 1974).

1.2.3.4 Light Trap with Provisions for Timely Automatic Separation of Insects

An automatic device for dividing and packaging light trap insect catches according to time intervals of as little as 5 minutes has been developed by Hartstack and Hollingsworth(1968) for overall insect survey. Insects collected in the gravity tape trap move from the killing chamber into a paddle wheel compartment which deposits them between sheets of rolled plastic film where they are retained until the package is opened for identification.

1.3 Uses of Light Trap

- Light traps are useful in studying the seasonal incidence of important insect pests and in evolving forecasting models.
- To survey the occurrence, distribution, abundance and flight pattern of various inset pests in a given locality.
- Gives an indication about brood emergence of insect pests for taking up appropriate control measures.
- Indicates migration of certain insect species.
- Can be effectively Controls certain pests light cotton bollworms, cucumber beetles, tobacco horn worm etc

1.4 Limitations

- Used to attract nocturnal insects.
- It should be fixed 2 - 3 feet above the vegetation level @ 5 traps / ha.
- Light trap catches will be less in full moon day × new moon day.

1.5 Target Pests

Leafhoppers, planthoppers, bugs, chaffer beetles, dung roller, rice leaf folder, stem borer, cotton boll worm, red hairy caterpillars, store grain pests etc.

1.6 Operation of Light Trap

Light traps also attract many beneficial and non target insects like parasitoids and predators. To avoid the loss of such insects, light traps have to be operated during peak period of activity of harmful insects. For example in rice ecosystem maximum numbers of insect pests are caught between 8pm - 12 midnight. The operations of light trap during this period will minimise the attraction of beneficial insects. In sorghum, the peak period of attraction of *Chilo partellus* is between 8 - 11 pm.

2. Sticky Trap

Sticky traps are generally used for catching the aerial insect population. Generally smaller size insects like leaf hoppers, white flies, aphids are trapped in this model. The sticky materials used include castor oil, vaseline, tangle foot etc.

This Sticky trap models include simple rectangular galvanized sheets, plastic jars or tin boxes, cylindrical tubes etc. over which the sticky adhesive materials is applied. Generally different colours are painted on the trap depending upon the preference by various insects to colours are used as coloured sticky traps. White coloured traps are most effective in attracting the pigeon pea fly, *Melanagromyza obtusa* yellow colour attract cotton white fly, *Bemesia tabaci,* cotton aphids, *Aphis gossypi* and green house white fly, *Trialeurodes vaporariorum* (Bhatnagar and Davies, 1979; Gillespie and Quiring, 1987). Models combining the sticky trap with water pan have also been developed to increase the insect catch. Sticky traps are generally used with pheromones (Mohan *et al.,* 1994).

2.1 Yellow Sticky Trap

Flying adult insects are getting attracted to yellow colour and they stick in the adhesive material of the trap. The traps can be made by using yellow polythene sheets or empty tin sheets in triangular shape or any desirable size and painted with yellow colour. The flags made from any one of them should be smeared with castor oil or grease and set in the different locations of the field at 30 cm above the crop level. The sticky material should be cleaned once in two days and replaced. The trap should be kept against the wind direction so as to attract more insects when they fly along the wind. All types of flying insects and sucking pests are attracted, in particular

Brinjal - White flies and Hoppers @ 5 traps/ acre.

Bhendi - White flies and Aphids @ 5 traps/ acre.

Cotton - Bollworms, White flies and Tobacco caterpillar @ 5 traps/ acre.

The insect attraction may depend upon the colour of the material that has been used. Here are some examples:

Red sticky trap – Apple maggots

White sticky trap – Weevils

Yellow water pan – BPH and WBPH

3. Pheromone Trap

Adult female insects that are ready for mating emit species specific chemical odours to attract males. Lab synthesized pheromones are specially packed and supplied as lures in rubber septa to be used along with suitable traps to deceive, attract and trap male insects. Traps were deployed at a height of 1M above average crop canopy @ 2-3 / acre. The distance between two traps was 40 ± 5 M at a distance, there is no attraction interference between two traps.

There are different types of pheromone traps are developed for collection of several insect pests. The types are given below.

3.1 Uni Trap

It is a funnel design to capture large numbers of insect pests. It can be used where large captures are anticipated. Pheromone lures are readily available are collected and installed in the lure basket. Uni traps are cost effective and durable for many season of continuous use. This is exclusively recommended for cut worms, army worms and other Lepidopterans.

3.2 Funnel Trap

A funnel is attached with a circular top having a provision for insertion of septa and the funnel terminating with a long polythene sleeve used as collection bag with a insecticide soaked cotton to kill the trapped insects. This model is most common among the farmers. This is recommended for Tobacco caterpillars, Boll worms etc.

3.3 Delta Type

This is also a popular type for monitoring of insect moths of diamond back moth and cotton pink boll worms. It is a triangular paper bowl with lure placed inside to capture pest insects in their sticky surface. Some insects may be hold at the entrance itself.

3.4 Bucket Type

In this type, the plastic bucket is wrapped with a coir material or gunny bag and there will be a hole made at the centre so as to make rhino beetles to get attracted towards the lure kept inside top of the closed bucket. The gunny bag wrapping is mainly to help the insect to hold the plastic bucket easily.

3.5 Uses

Emergence of the first insect of the season can be trapped. Where the pest species are coming from population trends from year to year can also be trapped.

3.6 Purpose

- *Monitoring* : To know the pest incidence and intensity in fields based on number of males caught, the timing and frequency of control measures can be determined.
- *Mass trapping* : As a result of mass trapping of insect males, the majority of females remain unmated and couldn't able to produce fertile eggs.
- *Mating disruption* : Wide spread application of synthetic pheromone in the field made the insects unable to locate their mates.

3.7 Available Lures: Spodo Lure – *Spodoptera litura*

Heli lure - *Helicoverpa armigera*

Pectino lure - *Pectinophora gossypiella*

Ervit lure and Ervin lure - Spotted boll worms

Ferro lure - Red palm weevil

Rhino lure - Rhinoceros beetle

4. Poison Bait Traps

Poison baits are food materials or chemicals mixed with poison or insecticide and used to trap insects. The effect is made to make the bait more attractive to insects than their natural food and also a smaller quantity should be able to attract the largest number of insects. Baits are used when for some reason spraying or dusting of insecticides is not practicable, when insects like hidden under the soil, inside the fruits and vegetables. There is different bait traps have been developed for attracting various insect pests of crops.

4.1 Fish Meal Trap

Moistened fishmeal is known to attract shoot flies attracting millet crops. It is widely used to trap sorghum shoot fly, *Atherigona soccata.* The trap developed at International Crops Research Institute for Semi Arid Tropics, Hyderabad is very effective in trapping shoot fly. It is made of plastic jar with perforations of 4mm diameter. The base of the jar is connected to a collection jar by a funnel. Moistened fishmeal is kept in a small plastic container fixed to the lid of the plastic jar. The lid also has a smaller container for keeping cotton soaked with insecticide DDVP to kill the attracted flies. The attracted and killed flies are collected in the collection jar. Collection jar can be detached from funnel to remove the flies. A low cost fishmeal trap has been developed at TNAU, Coimbatore. It is made of rectangular tin container with perforations of 4mm diameter on all sides. Moistened fishmeal is kept inside the container either in a polythene bag or in plastic container. The attracted flies are killed by keeping cotton soaked with DDVP.

4.2 Poison Balls

The food materials and chemicals are mixed together and made as bolls to attract insect pests. The poison balls can be prepared by mixing rice bran + jaggery + carbaryl + water @ 12.5 kg: 1.25 kg: 1.25 kg: 7.5 lit. and this is exclusively used to trap tobacco caterpillar, *Spodoptera litura* in various fields. The balls can be placed in different places at rows so as to make them equally distributed.

4.3 Chemical with Poison

In this category, chemicals can be mixed with any one of the insecticide and placed in the field to attract flies.

Ex.: Methyl Engenol: DDVP (1:1) - Fruit flies
Anisyl acetate: DDVP (1:1) - Fruit flies
Ammonia: Nuvan (1:1) - Wall nut husk fly

Geraniol: Carbaryl (1:1) – Fruit flies

Eugenol: DDVP (1:1) - Japanese beetle, Fruit flies

4.4 Other Baits

Molasses (or) toddy with 0.1% Nuvan can be used to attract coconut rhinoceros beetle and fruit sucking moths.

The following traps are mainly used to trap rodents in field as well storage godowns. These traps are functioning with any one of the baits without poison (insecticide) except poison bait mixture.

5. Poison Bait for Rodents

5.1 Single Dose Poison

Acute or single dose poison bait mixture can be prepared with Zinc phosphide (2g) + ground nut oil (3g) + popped corn / rice / dry fish (95g).

5.2 Multi Dose Poison

- *The dry bait* can be prepared by mixing flour (cereal or millets) 450g + edible oil 10g + sugar or jaggary1.5g and an anticoagulant, Warfarin 0.5% (25g) for effective rat control in the field as well as in store grain godown and warehouses.
- *Preparation of water soluble bait* by mixing 1 part of anticoagulant 19 parts of water (2.5g of anticoagulant dissolved in 475ml of water.
- Nowadays second generation anticoagulant namely Bromodiolone is available at 125g.

5.3 Limitations

- Before providing poison bait, keep the plain or non poisoned bait for 2-3 days as a pre bait to make the rats used to the food provided. Because rats are having bait shyness, it will not touch any new food item i.e. provided for them.
- Keep the bait with multi dose or chronic anticoagulant in small cups on the rat runs, dark places etc., where rats are frequently more.
- Replace the consumed bait daily, collect the rats which begin dying after 5 or 6 days and bury them.

- Use also water soluble bait in a shallow cups or plates in a number of places inside the godown for the rats to drink the poisoned liquid and killed.
- Discontinue the baiting or remove all the baited food and destroy as soon as the rat population is controlled.
- The bait should be kept 30 cm from bunds and 5 mts. gap between two places (bait stations) in the field. Several types of trapping devices are used to trap the rats in the field and in godown.

5.4 Thanjavur Bow Trap

The trap is made from woodern splits; the long stick is attached with a triangular bottom having a bow. The trap can be set up in the rice fields to catch the rats during day or night times. The bow softly hold by a small and thin piece of stick, the rats would get trapped and become dead when the rats moves to touch the food materials kept in both the sides. The trap is recommended @ 20 - 25 / acre. This is most common among the farmers of Tamilnadu and also works well in reducing the rat population in the field.

5.5 Back Break Trap

This trap is made of wood with some springs. The rats get trapped when the try touch the food materials kept in the iron rod connected with the release of strong spring and it hits back portion of the rats and made them to die immediately. The can be used any where even in the houses or godowns or in the fields also. This is very easy to handle and care should be taken to operate.

5.6 Sticky or Glue Trap

They are boards made of wood, hard- or cardboard covered with very sticky material. There are different types of glue available and they should be checked for suitability (stickiness, and usability in humid or dusty conditions) before large quantities are ordered. The boards are placed in the same way as traps, and normally there is no need for bait to attract rats. These traps should be checked daily, but are not regarded as very 'humane'. The use of glue traps (glue boards) should be limited. This type of trap consists of a sticky film of glue applied to a backing of cardboard, wood or plastic. These traps can fail when they get dirty, too hot or too cold. To help keep them free of dirt and moisture, glue trap covers can be used, or the traps can be placed in boxes with openings, in empty bait stations, etc. Nevertheless, savvy rodents will avoid them, vault over them or place debris on them to cover the sticky surface. Despite the many problems associated with their use, glue traps do catch the rodents and can play a part in an integrated rodent management programme.

5.7 Cage / Live Trap

Usually trap animals are unharmed so they can be released elsewhere (although sometimes the animals may injure themselves by trying to get out of the traps). Unlike the other traps, live traps can also be used outdoors and pose little risk to non target animals. Single catch live traps are rodent-sized cage traps of various styles. These traps capture the rat or mouse alive and unharmed, but the captured rodent must be dead with.

5.8 Electrocution Traps

They are a newer, cleaner, and more humane way to non-chemical control of the rats. The traps are baited with suitable bait (peanut butter, cheese, oats, salami, nesting materials, etc.) and placed in the rats' travel paths. When a rat enters the trap, it immediately receives a lethal electric shock. The traps that kill rats by electrocution are available on the market. These traps are considerably more expensive than the common snap trap and can be used safely and effectively only in limited situations. Like the snap traps, these traps catch only one rat at a time and then must be emptied.

5.9 Tribal Rat Trap

The tribal rat trap consists of medium sized, flat stone and three small pieces of elongated stick. The trap is set up in such a way that the rats would get trapped, under the falling granite stone when they try to touch the food material to eat. This happens as a result of release of the small piece of stick fitted with food stick. This can be made with in 5 - 10 minutes with the naturally available materials like stone and sticks. Unfortunately pieces of the stick fragile and will be damaged easily. It is known as Ralangal in the hills by the malayali tribes. The trap is set up in the field funds. This is in the use in the hills of Pachaimalai and Kalrayan hills of Tamil nadu (Kathirvelu, 2001; Kathirvelu and Narayanasamy, 2003).

5.10 Remodelled Tribal Rat Trap

The remodeled tribal rat trap was modified in a scientific way. It consisted of a flat concrete stone of 30×25 cm size and a thickness of 5 cm, an 'Y' like woodern stick made of strong neem wood with two arms each with 7.5 cm long and a central arm with 16 cm length. so as to keep their strangeness for long time (Narayanasamy and Kathirvelu, 2006).

5.11 Snap Traps

Snap traps should be set with the trigger end in the rats' travel paths, and only in areas inaccessible to children, pets, and non-target animals. Still, the best all-around trap for both mice and rats is the trusty snap trap. Modern snap traps have expanded plastic triggers proven to catch more rodents than older traps with smaller, metal triggers. Correct use of snap traps begins with proper placement. The snap traps should be placed along walls (rodents tend to run along walls), perpendicular to the wall, with the trigger end against the wall. They can also be placed in tandem (back-to-back), parallel to the wall so that rodents traveling in either direction will encounter the triggers. Snap traps should be placed at 10-foot intervals for mice and at 20-feet for rats.

6. Bait Stations

Bait stations are containers for placing poison bait and it protects the bait from rain, sunlight and other animals.

Ex: (a) Tin container with punctures - Fish meal trap
Mud pot - To keep molasses and fermenting sugars
Box type (Rectangular box with 7.5 cm dia. holes) - Rats
Bamboo type (Bamboo sticks one side closed with tin sheets) - Rats

7. Traps for Store Grain Pests (Mohan, 2009)

7.1 TNAU Insect Probe Trap

The trap is used for detecting / trapping in sects in stored grains. The basic components of a TNAU probe trap consist of three important parts: A main tube, insect trapping tube and a detachable cone at the bottom. Equispaced perforations of 2 mm diameter are made in the main tube. Insects love "air" and move towards air. This behaviour of the insect is exploited. This trap has to be kept in the grain like rice, wheat etc., vertically with the white plastic cone downside. The top red cap must be kept to the level of the grain. Insects will move towards air in the main tube and enter through the hole. Once the insect enters the hole it falls down into the detachable white cone at the bottom. Then there is no way to escape and the insects are trapped for ever. The white detachable cone can be unscrewed once in a week and the insects collected can be destroyed. This is considered as good mass trapping device when used at 2 - 3 numbers / 25 kg bin (28 cm dia and 39 cm length). They should be placed at top 6 inches of the grain, where the insect activity is seen during early period of storage. They can remove > 80% of the insects within 10 - 20 days (Mohan, 2009).

Target pests : Rhyzopertha dominica (F.), *Sitophilus oryzae* (L.) and *Tribolium castaneum* (Herbst)

7.2 TNAU Pit Fall Trap

Pitfall traps are used for capturing insects active on grain surface and in other layers of grain (Monitoring and mass trapping tool). Standard model of pitfall trap has 2 parts, perforated lid (2 mm (or) 3 mm) and a cone shaped bottom portion. Application of special coating with sticky material on the inner side of cone to hold trapped insects is necessary.

TNAU Model

TNAU model has perforated lid, cone shaped bottom which tapers into a funnel shaped trapping tube. Hence, sticky coating is dispensed. Commercial model is in plastic, simple and economical (cost per trap is Rs. 25/- only). This is easy to handle.

7.3 TNAU Two in One Trap

The probe trap containing the components namely the perforated tube, pitfall mechanism, a collection tube and the cone shaped pitfall trap with a perforated lid and the bottom tapering cone were combined as a single unit. Combination of probe and pitfall increase the trapping efficiency of insects. This is best suited for pulse beetles as they are seen only on grain surface wandering here and there. It does not require tedious procedures like coating the inner surface of pitfall cone with sticky materials before trapping to hold pulse beetles. Beetles are captured alive in this trap, which may facilitate release of pheromone and thereby attract more insects.

7.4 UV Light Trap for Store Grain Pests

The UV light trap mainly consists of a ultra-violet source (4 W germicidal lamp). The lamp produces ultra-violet rays of peak emission around 250 nanometer. The light is fitted at the centre of a funnel of 310 mm diameter at the top and 35 mm diameter at the bottom. The bottom end of the funnel is attached with a transparent plastic container for collecting the trapped insects. To hang the unit at desired points, three hooks have been provided at the periphery of the funnel. The unit is also provided with a tripod stand.

The UV light trap can be placed in food grain storage godowns at 1.5 m above ground level, preferably in places around warehouse corners, as it has been observed that the insect tends to move towards these places during the evening hours. The trap can be operated during the night hours. The light trap attracts stored product insects of paddy like lesser grain borer, (*Rhyzopertha dominica),* red flour beetle, (*Tribolium castaneum)* and saw toothed beetle, (*Oryzaephilus surinamensis)* in large numbers. Psocids which are of great nuisance in the godowns are also attracted in large numbers. Normally 2 numbers of UV light trap per 60 x 20 m (L x B) godown with 5 m height is suggested. The trap is ideal for use in the godowns meant for long term storage of grains, whenever infested stocks arrive in the godowns and during post fumigation periods to trap the resistant strains and left over insects to prevent build up of the pest populations. In godowns of frequent transactions the trap can be used for monitoring.

8. Others

8.1 Malaise Trap

The malaise trap is more elaborate it consists basically of an often fronted tent of cotton or nylon net, black or green in colour, the roof slopes upwards to the inner most corner at which there is an aperture leading to a trap. It was developed by Malaise (1937) as a collector tools; modified designs have been described Gressit and Gressit (1962), Townes (1972). The trap consists of a vertical net serving as a baffle end nets and a sloping canopy leading to a collecting device. This is loaded with a killing agent in order to stun and kill the insects that are later collected and preserved. Malaise traps are helpful in collecting a few insect species that are often difficult to collect otherwise.

8.2 Pitfall Trap

Pit fall trap helps to trap insects moving in the soil surface such as ground beetles, collembolans and spiders. These can be made by sinking glass jars or metal canes into the soil. It consists of a plastic funnel, opening into a plastic beaker containing kerosene supported inside a plastic jar. Some times the jar is loaded with baits or attracts to increase the efficiency of the collection.

9. Conclusion

The traps are major tools in the integrated pest management so as to avoid the resurgence, resistance with insects and residue problems in the eco system. This is also mainly used for mass trapping and monitoring of pests in a particular locality with low cost and it can be employed easily in the field to manage the pest population under control.

References

Bhatnagar, V. S. and J. C. Davies, 1979. Insects trap studies at ICRISAT center. Progress Report, Cropping Entomology 2. ICRISAT, Hyderabad, India.

Gillespie, D.R. and D. Quiting, 1987. Yellow sticky traps for detecting and monitoring green house whitefly (Homoptera: Aleyrodidae) adults on greenhouse tomator crops. *J. Econ. Entomol.* 80(4): 675-679.

Gressit, J.L. and M.K. Gressit, 1962. An improved Malasie trap. *Pacific Insects,* 4(1): 87-90.

Harding, W.C., J.C. Hartsock & G.G. Rohwer, 1966. Black light trop standards for general insect surveys. *Bull. Entomol. Soc. Amer.,* 12(1): 31-32.

Hienton, T.E. 1960. Response of insects to induced light paper presented in symposium organized by ARS USDA. Feb.,1960 at Maryland USA, 1-8.

Kathirvelu, C. 2001. Studies on tribal pest control practices of Tamilnadu, *M.Sc. (Ag.) Thesis,* Annamalai University, Annamalainagar.

Kathirvelu, C. and P. Narayanasamy, 2003. Rediscovery of tribal rat trap. *Invention intelligence,* 38(4): 193-194.

Mitchell, E.R., J.M., Stanley, J.C. Webb and Baumhover, 1974. Cylindrical electro grid traps. *Environ. Entomol.,* 3(1): 49-50.

Malaise, R. 1937. A new insect trap. *Entomolgisk Tidskrift,* 58: 148-160.

Mohan, S. 2009. Eco-friendly post harvest technologies for the management of stored grain insects at farm, home and warehouses – A step towards clean grain (insect free) revolution. Pp. 31-42. In: Narayanasamy, P., S.

Mohan and J.S. Awaknavar (Eds). Pest Management in store grains Satish Serial Publishing House, New Delhi, p. 272p.

Mohan, S., P.V. Subba Rao, and P.C. Sundara Babu, 1994. A new model trap for monitoring pigeon pea pod fly, *ICPN*, 1: 42.

Mohan, S., M.S. Venugopal, A. Janagarajan and M. Rangarajan, 1993. A simple trap for *Culicoides* sp. *TNAU News letter*, 23(4): p.4.

Narayanasamy, P. and C. Kathirvelu, 2006. Wapning tribal wisdom on pest control. Catterpillar Publications, Mariyappa Nagar, Chidambaram.

Parker, JR., A.L. Strand and H.L. Seamens, 1921. Pale western cut worm (*Porosagrotis orthogonia* Morr.). *J. Agr.Res.*, 22(6): 289-321.

Robinson, 1952. On the behaviour of night flying insects in the neighbourhood of bright sources of light. *Proc. R. Ent. Soc. Lond.* (A), 27: 13-21.

Taylor, L.R. and E.S. Brown, 1972. Effecting light trap design and illumination of samples of moths in the Kenya Highlands. *Bull. Entomol. Soc. Amer.*, 10(2): 13-15.

Townes, H. 1972. A light weight malaise trap. *Entomological News*, 83: 239-247.

❑❑❑

SECTION - VI

HOST PLANT RESISTANCE

26

Host Plant Resistance as Complement to Organic Pest Management

SELVANARAYANAN, V.
Professor, Department of Entomology, Faculty of Agriculture, Annamalai University
Annamalai Nagar – 608 002, Tamil Nadu, India

1. Introduction

Host plant resistance to insect pests is defined as the relative amount of heritable qualities possessed by the plant which influence the ultimate degree of damage done by the insect. In practical agricultural terms, an insect-resistant crop cultivar is one that yields more than a susceptible cultivar when confronted with insect pest invasion. The genetic diversity of many crop plants offers ample scope of exploitation of resistant traits. Cultivated varieties, wild lines, land races and other distantly related genera of crop plants are gathered, screened for resistance at the field and / or glasshouse conditions and promising resistant sources are exploited for large scale cultivation or future breeding programmes.

Insect-resistant crop varieties suppress insect pest abundance or elevate the damage tolerance level of the plants. In other words, insect-resistant plants alter the relationship of an insect pest with its plant host. The relationship between the insect and plant is influenced by the kind or mechanism of resistance, e.g. antixenosis (non-preference), antibiosis, or tolerance.

Resistance in a crop cultivar is exerted due to the presence of bio-physical or bio-chemical factors that result in any one of the above kinds of resistance.

Nonpreference in a cultivar may have either a bio-chemical or morphological basis. For example, cucurbitacins (tetracyclic terpenes) in cucubitaceous plants induce non-preference of spotted cucumber beetle, *Diabrotica undecimpunctata*. Hairiness or trichome density and many other structural characters in resistant crop plants offer morphological resistance to the infesting insect pests.

Antibiosis in resistant crop plants is mostly derived from allelochemicals and also few primary or secondary metabolites. For example, cyclic hydroxamic acid, (2,4-dihydroxy-7-methoxy-1,4-benzoxazine-3-one), DIMBOA in corn exerts antibiosis in European corn borer, *Ostrinia nubilalis* while aspargine in certain rice cultivars reduces the reproduction of brown planthopper, *Nilaparvata lugens*.

Unlike nonpreference and antibiosis, some of the components of tolerance mechanisms are general vigor, compensatory growth in infested plants, quick wound healing, changes in photosynthate partitioning, mechanical support in tissues and organs.

2. Tomato as a Model System

Keeping the above in mind, host plant resistance in tomato (*Lycopersicon* spp.) was studies as a model system. Among the various insect pests infesting tomatoes, the fruit borer, *Helicoverpa armigera* (Hubner) (Lepidoptera: Noctuidae) is more predominant and serious, whose young larvae feed on the leaves or flowers while matured instars feed on the fruits. Occasional boring into the vines is also encountered. In view of frequent outbreaks of *H. armigera* in various crops enabled by intensive crop production technologies, increased dependence on chemical pesticides leading to resistance development in addition to residues in freshly consumed commodities like tomato, identification and utilization of host plant resistance in tomato is an ecological safe and economically viable strategy in the integrated management of the fruit worm (Jallow *et al.*, 2004).

Many wild species of *Lycopersicon* such as *L. esculentum* var. *cerasiforme* (Dun.) (Chelliah and Srinivasan, 1983), *L. pimpinellifolium* (Jusl.) Mull. (Juvik *et al.*, 1982), *L. hirsutum* Hump. And Bonpl., *L. hirsutum f. glabratum* C.H. Mull. (Kashyap *et al.*, 1990), *L. cheesmanii*

(Bordat *et al.*, 1987) and *L. pennellii* (Goffreda *et al.*, 1988) are reported to be resistant to many insects and diseases.

Considering the above, tomato germplasm comprising 321 accessions, including wild species, land races, hybrids and cultivars (89% cultivated species, *L. esculentum*, 10% wild relative, *L. pimpinellifolium* and 1% suspected cross of this two) was screened for resistance against *H. armigera*, both under field and glasshouse conditions at Annamalai Nagar, Tamil Nadu, India from 1998 to 2000 and five accessions namely Varushanadu Local, PT 4287, Seijima Jeisei, Ac 238 and Roma were identified to be substantially resistant / tolerant (Selvanarayanan, 2000). Any attempt to capitalize on the resistance in tomato against the fruit worm, *H. armigera* should utilize the defense traits both in the foliage and fruits as the young larvae feed on tender foliage and flowers while older larvae feed on the fruits. On analyzing the various mechanisms and factors of resistance in these selected accessions in comparison with the susceptible check, I 979, lesser feeding preference of *H. armigera* towards the foliage and fruits was recorded. In contrast to this, these selected accessions recorded higher ovipositional preference but egg hatching got impaired considerably (Selvanarayanan and Narayanasamy, 2004). Larval feeding on the foliage and fruits of these accessions exerted pronounced antibiosis on *H. armigera* as evidenced by higher larval and pupal mortality; prolongation of larval and pupal tenure; decline in emergence and longevity of adults and also higher malformation in the emerging adults (Selvanarayanan and Narayanasamy, 2006).

Various bio-physical bases of resistance namely trichome density, calyx area, thickness and toughness of the fruit rind, fruit sepal thickness, seed: pulp ratio were examined among which, trichome density was found to exert a significant negative correlation with larval feeding. Trichomes on the foliage of *Lycopersicon* spp. Have been categorized as types I to VII, among which types I, IV, VI and VII are glandular and types II, III and V are non-glandular. In this study, the selected accessions had two non-glandular types (III and V) and three glandular types (I, VI, VII).

On estimating the content of various biochemical factors of resistance namely, reducing sugars, non-reducing sugars, total sugars, amino acids in the foliage and fruits; acidity, ascorbic acid, lycopene in the fruits; total phenols, O. D. phenols and chlorogenic acid in the foliage, it was observed that phenol content of the foliage and acidity of the fruits exerted a significant negative correlation with larval feeding. Besides these, the type VI glandular trichomes of *L. hirsutum f. glabratum* were reported to possess methyl ketones such as 2-tridecanone and 2-undecanone, while that of *L. pennelli* possess acyl sugars.

Realizing the value of these accessions selected, they were subjected to intercrossing by conventional hybridization (emasculation and pollination), which yielded three viable hybrids namely Hybrid 1 (Varushanadu Local x Ac 238), Hybrid (Varushanadu Local x Roma) and Hybrid 3 (Ac 238 x Roma). The resistance potentials of the selected accessions and their hybrid derivatives were studied both in the glasshouse and field, in comparison with a susceptible check against fruit worm, *H. armigera* (Dhakshinamoorthy, 2002), serpentine leaf miner, *Liriomyza trifolii* Blan., whitefly, *Bemisia tabaci* Genn., and leaf caterpillar, *Spodoptera litura* Fab. (Muthukumaran, 2004). Though the hybrids recorded tolerance to many of the pests at the field level, a wider variation was observed among them suggesting the possibility of segregation of the defense traits. This was further evidenced by the variation observed in the trichome density and phenol content between the F1 and F2 generations.

3. Plant Resistance under Organic Insect Pest Management

Plant resistance as the sole or primary tactic: Resistant varieties are first line of defense in cases where pesticides can not be used or other alternative controls are not available.

Plant resistance integrated with other tactics: Resistant cultivars are compatible with chemical control, cultural control, behavioral control, and in most cases is also compatible with biological control strategies.

With the increasing awareness and need for organic pest management techniques, host plant resistance may play a pivotal role on which all the other techniques may be employed in an integrated manner.

3.1 Plant Resistance as a Compliment to Cultural Control

Cultural control involves all the techniques (i) that make the environment less favourable to the pest and (ii) that make the environment more favourable to the pest's natural enemies. Plant resistance in combination with cultural control can effectively reduce the need for chemical pesticides.

3.2 Synchrony of Plant and Herbivore Phonologies

The plant and pest phonologies are carefully studied and either avoiding or overlapping the phonologies may be practiced. The temporal alterations aim at reducing or avoiding the pest incidence. The best example is the combination of plant resistance and short - growth-duration cotton varieties which effected an efficient management of boll weevil and pink bollworm in USA.

3.3 Vegetational Diversity, Variety Mixtures and Trap Cropping

Population densities of insect pests are frequently lower in polyculture habitats which are ecologically complex, supporting the natural enemies by providing more resources and offering less concentrated resources for specialist herbivores. In a similar way, crop cultivars with varying degrees of resistance may favour natural enemies and thereby unfavourable to herbivores. The concept of trap cropping has been widely exploited in many crop ecosystems. Crop stands are manipulated in time and space, so that attractive host plants are offered at the critical time of the pest's phenology. Such tactics lead to the concentration of the pests in the trap crop and thereby enabling least or no insecticidal application in the target crop.

4. Plant Resistance and Biological Control

Integrating biological control and resistant varieties has been successful in many cases. In general, insects feeding on resistant host plants commonly experience retarded growth and an extended developmental period (antibiosis). Such poorly developed insects are more vulnerable to natural enemies. In addition to this, in certain cases, the plant secondary compounds that impart resistance are compatible with the performance of natural enemies. The best example for such complimentary interaction was reported with resistant varieties of barley, the green bug, *Schizaphis graminum* (Rondani) and the parasitoid, *Lysiphes testaceipes* (Cresson). Similarly, low level of antibiosis in moderately resistance tobacco plants results in extension of early instar stage of *Heliothis virescens* thereby making it more parasitized by *Campoletis sonorensis*

However, it should be cautioned that resistant varieties are not always compatible with biological control programs. For instance, resistant tobacco varieties with glandular and hooked trichomes have been shown to severely limit egg parasitization of tobacco hornworm eggs by *Telenomus sphingus* and *Trichogramma minutum.* Peter *et al.* (1995) have given a lucid review of the role of plant trichomes in insect resistance and in fact, tomato varieties with denser trichomes though resist insect activity, exert a negative influence on the existence, development and efficiency of parasitic fauna (Farrar and Kennedy, 1993).

5. Conclusion

Plant resistance has many advantages as a primary tactic in insect pest management. The important advantages of plant resistance are (i) Specificity; (ii) Persistence; (iii) Compatibility; (iv) Eco-friendliness; (v) Ease of adoption; and (vi) Cumulative effectiveness. In addition to the above and unlike biological or insecticidal control tactics, plant resistance is least influenced by weather parameters.

But a higher level of resistance or immunity may not always be attainable and it is not needed also. A resistant plant variety that reduces the insect pest population by 50% in each generation is ideal for reducing the insect population to below the economic injury level within a few generations (Painter, 1958). A moderate level of resistance or tolerance does not place any selective pressure on insect populations.

Although these advantages outweigh the disadvantages, nevertheless there are some important limitations to this approach. The time required for gathering, identifying and developing resistant varieties from the vast gene pool of many crop plants may be as long as 15 to 20 years. This approach necessitates great deal of expertise, vast resources and relatively long term funding. Further, some of the resistant traits in crop plants require diversion of resources and often there exists a negative correlation between the potential yield and level of resistance. In addition to the above, conflicts of resistant characters between different pests, evolution of biotypes, incompatibility of certain resistant traits with natural enemies or insecticides; poor agronomic and/ or palatability characters of some of the resistant varieties are the disadvantages that mar the use of plant resistance for insect pest management.

The future thrusts of plant resistance are focused to further exploit the resistant traits by breeding programs employing marker aided selection and identifying the respective QTL. Induction of resistance may also be attempted by suitable elicitors such as nutrient amendment. Novel techniques such as embryo rescue method or somoclonal variation may also be utilized if wild varieties are to be exploited.

References

Bordat, D., H.Laterrot, M. Renand and A. Morethi. 1987. Phenomena of *Lycopersicon cheesmanii* Riley resistance to *Liriomyza trifolii* (Burgess) and *L. sativae* Blanchard (Diptera: Agromyzidae). *Agron. Trop.*, 42: 43-46.

Chelliah, S. and K. Srinivasan. 1983. Resistance in bhendi, brinjal and tomato to major insect and mite pests. Pp. 43-44. *In:* Proc. National Seminar on Breeding Crop Plants for Resistance to Pests and Diseases, May 25-27, 1983, Coimbatore, India.

Dhakshinamoorthy, G. 2002. Studies on resistance of tomato against fruit borer, Helicoverpa armigera (Hubner). M.Sc. (Ag.) Thesis, Annamalai University, India.

Farrar, R.R., Jr. and G.G. Kennedy. 1993. Field cage performance of two tachinid parasitoids of the tomato fruit worm on insect resistant and susceptible tomato lines. *Entomol. Exp. Appl.*, 67: 73-78.

Goffreda, J.C., M.A. Mutschler and W.M. Tingey. 1988. Feeding behaviour of potato aphid affected by glandular trichomes of wild tomato. *Entomol. Exp. Appl.*, 48: 101-107.

Jallow, M.F.A., Paul Cunningham, J., Zalucki, M.P., 2004. Intra-specific variation for host plant use in *Helicoverpa armigera* (Hubner) (Lepidoptera: Noctuidae): implications for management. *Crop Prot.*, 23, 955-964.

Kashyap, R.K., M.K. Banerjee, G. Kalloo and A.N. Verma. 1990. Survival and development of fruit borer, *Heliothis armigera* (Hubner), (Lepidoptera : Noctuidae) on *Lycopersicon* spp. *Insect Sci. Appl.*, 11: 877-881.

Muthukumaran, N. 2004. Resistance of Lycopersicon accessions and their hybrid derivatives against certain insect pests. M.Sc. (Ag.) Thesis, Annamalai University, India, p.94.

Painter, R.H. 1958. Resistance of plants to insects. *Annu. Rev. Entomol.*, 3: 267-290.

Peter, A.J., T.G. Shanower and J. Romies. 1995. The role of plant trichomes in insect resistance: A selective review. *Phytophaga*, 7: 41-63.

Selvanarayanan, V. 2000. Host plant resistance in tomato against the fruit borer, *Helicoverpa armigera* (Hubner). *Ph.D. Thesis*, Annamalai University, Tamil Nadu, India.

Selvanarayanan, V. and P. Narayanasamy, 2004. Antixenosis resistance in tomato against fruit borer, *Helicoverpa armigera* (Hubner). *Int. J. Tropical Insect Sci.*, 24(3): 201-206.

Selvanarayanan, V. and P. Narayanasamy, 2006. Factors of resistance in tomato accessions against the fruit worm, *Helicoverpa armigera* (Hubner). *Crop Prot.*, 25: 1075-1079.

❑❑❑

SECTION - VII

FLYASH IN PEST CONTROL

27

Flyash : Potential in Organic Pest Control

NARAYANASAMY, P.
Dean, Faculty of Agricultural & Professor and Head, Department of Entomology, Faculty of Agriculture, Annamalai University, Annamalainagar 608 002, Tamilnadu, India

1. Introduction

Pesticide consumption in India has reached a point wherein residues of the pesticides remain forever in soils, underground water, run-off water on earth, food grains, vegetables, tubers animal flesh, cattle feed and mother's milk. This is evident from the recent skirmishes on soft drinks due to prepared with underground water polluted with DDT, the first known insecticide used for nearly two and half decades and banned later from synthesis and application. This teething problem warrants an appropriate alternative to the hazardous pesticides in the form of substances of biological and geological origin. Biological pesticides therefore, have tremendous value in plant protection in agriculture. Similarly coal and lignite unearthed from the underground are utilized immensely in the power generation and the residue accruing out of them at the thermal plants which is named as 'flyash', a material of concern from the point of view of its disposal at the production site (Ahmed and Stoll, 1996).

2. Flyash Scenario (Narayanasamy, 1994)

The flyash owing to its composition of various chemical elements especially silica and an array of elements has been now designated to be a manure for crop plants and interestingly, the flyash has been discovered to be a pesticide to check varieties of pests hampering crop plants, which only led to reduced application of chemical pesticides.

The scenario of disposal of huge quantity of the flyash has evinced interest from the cement industries, brick kilns, highways and agriculture sectors. Agriculture in particular has been observed to be an area for profitable use of the flyash for enhancing crop yields, improving soil nutrient status and crop protection from the pest depredations.

Flyash a residue accumulating at the sites of coal and lignite based thermal Power Plants has because a problem of disposal owing to productive enormously and subquent Environmental Pollution. At present approximately 290 million rabi of coal is consumed per annum leading to production of 112 million rabi of flyash annually. The generation of the flyash is expected to exceed 175 million tones/year by 2012 and 225 million tones by 2017. Such large volume of the flyash will occupy large area of land and pose a great threat to the environment and the agricultural ecosystem as well.

In this context, it is hereby indicated that the flyash has been utilized for gainful purposes in different sectors of life including agriculture. Application of the flyash in Agriculture has been positive for utilities such as soil conditioning, plant nutritioning, plant protection, etc.,

Pesticidal value of the flyash has been discovered by us at the Department of Entomology, Annamalai University. Flyashes of both coal and lignite have been of value in this regard. Besides, the flyash has been found to be serving as a carrier in the formulations of pesticides. Application of the flyash as a pesticide has been standardized for various crops such as, rice, vegetables namely, brinjal, bhendi, tomato, cucurbits, crucifers and fruits like, mango and citrus. Mechanisms of action of the flyash in insects have been well elucidated. Based on the findings obtained over a period of 15 years of research, packages of recommendations of the flyash have been formulated for a number of crops for adoption by the farmers.

3. Research Background on Flyash

Considering content of Silica (as SiO_2) our research commenced on the flyash in the year 1987 and subsequently a three year research project titled, "Studies on Utility of Lignite flyash as an Insecticide and an adjuvant in pesticide formulations" funded by the TamilNadu State Council for Science and Technology, Govt. of TamilNadu, Chennai was undertaken. This study for the first time, has brought to light insecticidal nature of the flyash in tackling pests problems in crops like rice. Also the flyash has been found to be an efficient carrier in insecticide formulations like Dust, Wettable Powder and Granule. Certain flyash based insecticides developed by us were found efficient against pests of rice. This led to furthering our research on the application of the flyash of coal and lignite in other crops like brinjal, bhendi tomato, cucurbits, crucifers and fruits like mango and citrus. The research accomplishments emboldened us to organize for the first time in India National Seminar on Use of Lignite fly ash in Agriculture during 1995 and subsequently another symposium in the year 2001 with strong bondage and support of the Flyash Mission, TIFAC, Department of Science and Technology, DST, Govt. of India and Neyveli Lignite Corporation Ltd., Neyveli, Tamil Nadu. This enabled us to collate researches made so far on the application of flyash in agriculture.

4. Collection, Transformation and Application of Flyash as Pesticide

Lignite Flyash was collected from the neighbouring Thermal power plant of Neyveli Lignite Corporation Ltd., Neyveli and Mettur Thermal Station, Tamil Nadu. The flyashes of coal and lignite were processed and used as dusts capable of being applied with knapsack rotary and plunger dusters.

Firstly Flyash 100% Dust was dusted on the plants and before alighting of on the plants insects were allowed to feed on them. Also it was applied on body of the caterpillar pests to find its effects if any, on the body biochemicals. Performance of the flyash in field condition was also ascertained in a number of crops including rice.

Secondly Flyash was used as a carrier in the synthesis of certain insecticides viz., B.H.C. 10% Dust, B.H.C. 50% WDP, Malathion 25% WDP and Carbofuran 3% WDG in the place of the conventional carriers as an attempt to find its suitability and reduction in the cost of the finished product namely 'Flyash based insecticides'.

Flyash dust and Flyash based pesticides are designated as "Fly ash pesticides".

5. Flyash Acts as an Insecticide (Narayanasamy, 1997)

- Lignite Flyash (LFA) has been discovered as a potential dust insecticide against various pests of crops like rice, vegetables and fruits.
- In rice flyash applied @ 16 kg/acre killed leaf eating pests like leaffolder, yellow hairy caterpillar, small grasshopper and sucking pests like brown planthopper, green leafhopper, earhear bug and brown bug.
- Enhanced doses of LFA applied to rice plants were found detrimental to the brown planthopper, a major sucking pest of national importance in rice. Application of LFA 2 times at 30, 50 days after transplanting was found appropriate to check the pest attack.
- Application of Lignite Flyash (LFA) and Coal Lignite Flyash (CLFA) showed their insecticidal nature to pests of vegetables like brinjal fruitborer, Epilachna beetle, bhendi fruit borer, leaf caterpillars and tomato fruit borer in the field and laboratory conditions.
- When larval pests were forced - fed with LFA, there was 100 % mortality of pupal stage of bhendi fruit borer, Earias vitella. In those cases from which adult moths emerged, the adults were malformed with physical deformities that they were unfit for further life activities. This interesting phenomenon will culminate in deciphering of the pest population in the subsequent cropping season.
- Bhendi leafhopper, Amrasca devastans when fed with liquid diet containing 2.5% extract of flyash succumbed to death. Similar trend will be useful against other related sucking pests of vegetables when the flyash is recommended for soil application.
- Larvae of cabbage pests like diamond back moth and leaf webber, bittergourd leaf beetle (Epilachna), mango leaf webber and citrus butterfly suffered high mortality due to LFA treatment in their feed material both in field and laboratory conditions.
- It can be used as termitcide effectively according our studies

- Bio-efficacy of flyash based herbal pesticides has been attempted for the first time and their efficacy on certain insect groups is reported. All the eight flyash (FA) based herbal pesticides had showed their efficacy in thwarting various groups of pests infesting rice and vegetables thus revealing them to be potential biopesticides. Of all the treatments, FA + Turmeric 10% Dust and FA + Neem Seed Kernel 10% Dust were found to be the most potential biopesticides against all the test insects including Epilachna on brinjal and Spodoptera on okra followed by FA + Vitex 10% Dust and FA + Eucalyptus 10% Dust and FA + Ocimum 10% Dust.

5.1 Mechanisms of Action of Flyash in Insects (Narayanasamy, 1994)

5.1.1 Flyash Causes Loss of Body Moisture in Insects

Flyash, when applied topically on rice pests such as leaf folder, horned caterpillar, skipper, yellow hairy caterpillar and cutworm had caused loss of body moisture, body fat and proteins, before death.

LFA removed epicuticle in the bodies of pests like rice yellow hairy caterpillar, tomato fruit borer (Helicoverpa armigera) and tobacco caterpillar (Spodoptera litura) and finally the pests succumbed to death.

5.1.2 Flyash Causes Histological Disfiguration in Insects

Flyash particles consumed by the insects along with their feed had caused cellular disintegration and large scale damage was done to vital organs of the body. This feature was tracked in many pests of vegetables, mango and citrus.

5.1.3 Flyash Disrupts Secretion of Enzymes

Flyash had impaired production of body enzymes like proteases, pectinases, etc. leading to collapse of physiological activities in insect pests like.

5.1.4 Flyash Acts as a Carrier in Pesticide Formulations

In lieu of common carriers being used in the formulations of the pesticides the flyash was put into use and insecticides like BHC 10% Dust, Malathion 25% WDP, BHC 50% WP and Carbofuran 3% WDG were developed. All these "flyash based pesticides" excelled the commercial pesticides in controlling the pests infesting rice like leaf folder, yellow hairy caterpillar, small grasshopper, brown planthopper, green leafhopper, earhead bug and coccinellid beetle in field conditions.

5.1.5 Flyash Induces Resistance in Rice Plant to Pests

When flyash was incorporated into soil, rice plants got more layers of silicified cells which made their culms and leaf lamina and leaf sheath thick and pests like leaffolder, skipper, horned caterpillar, yellow hairy caterpillar, cutworm and grass hopper could not feed continuously due to wearing of their feeding organs. This is a case of induced resistance in rice plant to the pests.

5.1.6 Flyash Mixes with Conventional Insecticides

LFA was found amenable for mixing with the commercial dust formulations like BHC 10% Dust and Carbaryl 5% Dust and efficient at 25% proportion to control the rice pests.

5.1.7 Flyash Protects Beneficial Organisms

Flyash was found safe to various forms of beneficial fauna found in rice ecosystems like spiders, predatory crickets, damselflies and parasites and fungal pathogens surviving on the pest forms.

5.1.8 Flyash is Safe

Flyash was found safe to albino rats, fishes and earthworms and beneficial insects like silk worms and honeybees through various modes of application.

Based on the findings, applications of the flyash have been standardized and recommendations made for rice, vegetables and fruits.

5.1.9 Other Utilities of Flyash in Pest Control Arena

- Flyash is useful as a potential insecticide for all kinds of chewing insect pests in rice, vegetables, etc.
- Flyash is useful as grain protectant against feeding of pests under storage condition.
- Flyash is useful as a potential carrier in certain insecticide formulations like Dust, Wettable Powder and Granule.
- Flyash dust is compatible with other insecticide Dusts by causing enhanced control of variety of pests of rice.

- Flyash is safe to the biological pest control fauna of rice fields such as predators (spiders, damsel flies, crickets) parasitoids and entomozfungi, productive insects like mulberry silkworms and honey bees and beneficial animals like frogs, fishes and earthworms.
- Flyash shows itself to be a good additive in the preparation of the Mycoinsecticide 70% WP, a fungal biopesticide developed to curb certain key pests in rice.

6. Flyash as a Carrier in Insecticide formulation (Narayanasamy, 1997)

Further, the LFA has been found to be an active carrier in certain insecticide formulations like dust, wettable powder and granule. Insecticides like B.H.C. 10% dust, B.H.C. 50% W.D.P., Malathion 25% W.D.P. and Carbofuran 3% W.D.G. were synthesized using fly ash as carrier and other ingredients as per ISI specifications laid down for the formulations. The fly ash based pesticides (FABP) when evaluated in field against pests of rice revealed that all of them excelled the commercial chemicals of the respective group in controlling the target pests. Economics of FABPs indicated that cost of preparing dust and granule with fly ash was calculated to be 2 – 3 times less than that of the respective commercial chemicals while fly ash wettable powder was 1.5 times less. Cost of 1kg 100% Insecticidal Dust is Rs. 1/- only.

Recently, eight " Flyash Herbal pesticides" were synthesized certain important pesticidal plant species like, Neem, Indian privet, Adhatoda, Pepper, Turmeric, Chilli, Ginger etc. and tested for their quality. These bio-pesticides were detected to be excellent pesticides in crops like rice, oilseeds, vegetables and store grains, these products will go in the name of "Flyash Biopesticides" and will have greater role in the management of the current pesticides imbroglio besides alleviating pesticide pollution from the agricultural ecosystem. It is therefore felt that the flyash might replace the chemical pesticides in the long run.

7. Recommendations of Flyash for Various Crops

Following are the recommendations of using the flyash for pest control in certain crops.

• Flyash 100% Dust	@ 40 kg/ha at 30, 50 & 70 DAT	For pests of rice like BPH, GLH, leaf folder, grasshopper, earhead bug and leaf beetles
• Flyash 100% Dust	@ 40 kg/ha at 20, 35, 50, 65, 80 DAS	For pests of bhendi like stem and fruit borer, aphids, leafhopper and leaf roller
• Flyash 100% Dust	@ 40 kg/ha at 35, 50, 65, 80 DAT	For pests of brinjal like shoot and fruit borer and spotted beetle
• Flyash 100% Dust	@ 40 kg/ha at 30, 45, 60, 75 DAT	For pests of tomato like fruit borer and leaf miner

8. Economics of Disposing the Flyash as a Pesticide (Zeaudeen *et al.*, 1995)

Cost of 1 kg Flyash 100% Dust will be 3-5 times cheaper than the commercial insecticide like Endosulfan 5% Dust.

In the case of the Flyash based insecticides such as BHC 10% Dust, BHC 50% WDP, Malathion 25% WDP and Carbofuran 3% WDG, the cost is nearly 2-3 times cheaper than those of the respective commercial insecticide. It becomes clear that the cost of the pesticides developed with the flyash as a carrier is significantly reduced. Plant protection cost borne by the farmer is also declined thus increasing net profit from the crop produce.

Considering the dose of the flyash at the rate of 10 Kg/ acre standardized through our studies, about 12 million rabi of flyash might be disposed to cover 150 million hectares of agricultural land and management of the flyash at the thermal power plants is minimized. Such large dose of flyash use in agriculture is interesting and economical as it might fetch us a revenue of Rs. 1200 Billion annually when the price of the flyash is fixed at a rabic Rs 1/ Kilogram as against Rs. 5 / Kilo for a dust insecticide sold right now. As a carrier in the formulations of the pesticide especially as dust and wettable powder, the flyash may get us about Rs 15 million for marketing approximately 30,000 tonnes of flyash pesticide per year.

Efforts have now been initiated to develop "Flyash Herbal pesticides" and "Flyash Mycoinsecticides". This involves certain potential pesticidal plant species like, Neem, Indian privet, Adhatoda, Pepper, Turmeric, Chilli, Ginger, etc. These products will go in the name of "Flyash Biopesticides" and will have greater role in the management of the current pesticides imbroglio besides alleviating pesticide pollution from the agricultural ecosystem. It is therefore felt that the flyash might replace the chemical pesticides in the long run.

9. Conclusion

The flyash is therefore a universally accepted wealth from waste. The need of the hour is to create social awareness regarding the bulk utilization of the flyash for a wide variety of applications. Disposal of the flyash in large volumes will incidentally save us from pollution hazards. Creation of necessary infrastructure facilities and marketing sector for the utilization of the flyash products is a Sine qua non at the present juncture.

From the experience gained, the flyash is acting as potential dust insecticide to tackle the pest problems in the rice, cauliflower, tomato, brinjal and bhendi and is also a carrier in the pesticide formulations. Attempts are to be ventured into delivering the flyash products commercially into the hands of the farming community. The flyash appears promising as a gift to all concerned in the new millennium.

Acknowledgement

I thank sincerely authorities of Annamalai University for enabling me to do the above pioneering work on the Flyash and for all facilities provided. Also I extend my gratitude to the Tamil Nadu State Council for Science and Technology, Government of Tamil Nadu, Chennai for sponsoring a project on the Flyash.

I take this opportunity to extend my grateful regards to Dr. Vimal Kumar, Scientist & Head Flyash Unit, Dept. of Science and Technology, Govt. of India for continued support rendered in the conduct of National Seminars on flyash at Annamalai University and for his interest and for encouragement for flyash research through DST funded project.

References

Ahmad, S. and G. Stoll, 1996. Biopesticides, In: Biotechnology Building on Farmers Knowledge, Eds. Joske Bunders, Berts Haves Kort and win Hiemsha, Mac Millan Education Ltd., London, pp. 52-79.

Narayanasamy, P. 1994. Studies on the utility of lignite flyash as an insecticide and an adjuvant in insecticide formulations, Final Project Report Tamil Nadu State Council for Science and Technology, Govt. of Tamilnadu, Chennai, pp. 144.

Narayanasamy, P. 1997. Development and use of flyash pesticides, In: Fly Ash in Agriculture, Proc Natl. Sem. On use of Lignite Fly Ash in Agriculture, Ed. P. Narayanasamy, Faculty of Agriculture, Annamalai University, pp.1-4.

Zeaudeen, P., R. Sita Devi and P. Narayanasamy, 1995. Profitabillity of AU-1 Mycoinsecticides 70% WP in use against rice brown planthopper, In: Proc. UGC Interdisciplinary Colloquium on Biopesticides in Indian Agricuture, Faculty of Agriculure, Annamalai University, Annamalainagar, India. Pp. 40-45.

❑❑❑

SECTION - VIII

PHYTONEMATODES

28

Potential of Biocontrol Agents in the Management of Nematode Pests of Crops

JONATHAN*, E.I., S. PRABHU, K. DEVRAJAN and P. SENTHILKUMAR
Department of Nematology, Tamil Nadu Agricultural University, Coimbatore - 641 003.
*Director, Centre for Plant Protection Studies, TNAU, Coimbatore-641 003

1. Introduction

In the recent past there have been several developments globally in the field of Nematology. Several nematicides have been withdrawn from the market due to increasing awareness about environmental risks thus necessitating it to look for alternative methods. Another important development that took place was demonstration of effective suppression of nematodes (cyst and root-knot) in specific soils rich in antagonists (fungi and bacteria) which opened new area of feasible nematode control, through bioagents. This inturn has lead to the availability of commercial formulations which are now being used to a limited extent in the West. However, the Indian scenario is yet to visualize commercial formulations of nematode antagonists in the market for the use by farmers.

Several organisms, such as fungi, bacteria, viruses, predatory nematodes, insects, mites and some invertebrates have been found to parasitize or prey nematodes and among these, some are potential biocontrol agents. The most widely studied organisms are antagonistic fungi, bacteria and predatory nematodes are furnished in Table 1.

Table 1: Common characteristics of some potential biological control agents against plant parasitic nematodes

Type of Bioagent	Mode of action	Advantages	Limitations
Trichoderma viride	Hyphal penetration	Easily produced *in vitro*, rhizosphere competent; wide host range; attacks eggs of several nematode spp.; treatment of planting material effective.	Efficient saprophytic competitor; Only seed treatment is effective;
Paecilomyces lilacinus	Hyphal penetration	Easily produced *in vitro*, rhizosphere competent; wide host range; attacks eggs of several nematode spp.; treatment of planting material effective.	Require high soil temperatures; variable efficacy in range of conditions; large inoculum required; some isolates are pathogenic to humans.
Pochonia (Verticillium) chlamydosporium	Hyphal penetration	Easily produced *in vitro*; rhizosphere competent; wide host range; attacks eggs of several nematode spp.; resistant resting spores produced; survives throughout season.	Seed treatment ineffective; efficacy depends on nematode spp., density and plant host.
Trapping fungi	Traps produced on modified mycelium which give rise to infective trophic hyphae.	Easily produced *in vitro*; some spp. Are rhizosphere competent; wide host range.	Difficult to ensure trapping activity coinciding with activity of cyst and root-knot nematodes; often do not produce resting spores.
Arbuscular Mycorrhiza (AM)	Competition in roots and modification of roots	Include agents with potential to control migratory endoparasitic nematodes in roots; may improve plant growth even in absence of nematodes; can be mass produced and formulated; could be applied to seeds or transplant material; may also reduce fungal rots.	Non-mycorrhizal fungi may be difficult to register as they are closely related to plant pathogens; efficacy affected by plant cultivar and crop rotation. Crucifers can reduce survival.
Pasteuria spp.	Adhesive spores	Most isolates highly virulent; infective spores resistant to drying; good shelf-life; reduce infecti-vity of nematodes and fecundity.	*In vitro* production not possible; isolates are highly specific; no proliferation in soil in the absence of nematodes.
Rhizobacteria	Toxins, antibiotics and/or	Easy to culture *in vitro*; can be applied as seed dressers; reduce plant damage.	Effective for relatively short period; activity affected by crop

2. Potential of *Trichoderma viride* as Bioagent

Trichoderma are one of a small group of beneficial fungi present in nearly all soils and has diverse habitats. They have proven commercially viable as a biological control agent.

There are many species of *Trichoderma* with different characteristics. For example, *T. harzianum* is tolerant to stress imposed by nutrient scarcity. Often they are antagonistic towards one another. *Trichoderma* is able to grow in soils having a wide range of pH (2.5 - 9.5) although most them prefer a moderately acidic environment. The species that prefer the more acidic soils are usually regarded as having a more stress tolerant growth habit and are less aggressive.

They readily colonize the root surface and cortex of higher plants. The best strains of *T. viride* will colonize root surfaces even when roots are a meter or more below the soil surface. Some strains can persist upto 18 months after application.

2.1 Mechanism of Action Against Phytonematodes

Mycoparasitism

Appresoria of *Trichoderma harziamum* were observed on the surface of cysts and eggs of *Globodera rostochiensis*. The fungus digests the cyst surface forming a channel in which it enter the cyst wall and grow subcuticularly. The cyst wall and egg shells of nematodes are barriers for predators, but *Trichoderma* is able to penetrate through such barriers.

The fungus has been reported to secrete many lytic. Among the Chitinases of *Trichoderma viride* help in parasitizing of eggs *Meloidogyne T. longibrachiatum* produces acetic acid which has nematicidal property.

Antibiosis

Sivasithamparam and Ghisalberti (1998) listed 43 substances which are produced by *Trichoderma* spp. Which have both nema static and produces antibiotic *viz.* trichodermim, dermadin, trichoviridin and sesquiterpene heptalic acid. The UV mutants of *T. harzianum* produced trichozianines, which are hydrophobic peptides that interact with rabicapids membrane and induced membrane permeability. Alkylpyrones, isonitriles, polyketides, peptaibols, diketopiperazines and steroids were frequently been associated with biocontrol activities of some species and strains of *Trichoderma*.

Competition

Competition for space or nutrients has long been considered one of the "classical" mechanisms of biocontrol of *Trichoderma* spp. They have very high competitive saprophytic ability which colonize the rhizosphere and roots which reduces the feeding sites for nematodes.

Tolerance to Biotic and Abiotic Stresses

Another possible mechanism recently gaining credence is tolerance to stress through enhanced root and plant development. Enhanced rooting by *T. hazianum* also induces tolerance to pests that it does not directly control. The larger root systems of plants colonized by *Trichoderma* were better able to withstand the damaging effects of the nematodes.

Induced Resistance

Trichoderma viride was observed to induce systemic resistance against root-knot nematode *Meloidogyne incognita* in green gram. Enzymes such as peroxidase, polyphenoloxidase, phenylalanine ammonia lyase, catalase and chitinase were induced more in *T. viride* treated plants compared to untreated plants. Induction of these active defence mechanisms determine the ability of a plant to survive the nematode attack.

Paecilomyces lilacinus. A potential antagonist of plant nematodes

Biological Mode of Action

Paecilomyces lilicinus has been reported to parasitize eggs of many sedentary endoparasitic nematodes. The infection process begins with growth of the fungal hyphae in the gelatinous matrix; eventually the eggs of *Meloidogyne, Tylenchulus* or *Nacobbus* species are engulfed by the mycelial network, which becomes prostrate and spirals over the smooth egg surface. The proliferated hyphal branches penetrate the eggs. In cysts of *Globodera* species the fungus penetrates the vulva or the broken and exposed neck region. After entering the cysts the fungus grows saprophytically on the mucilaginous body content surrounding the eggs during or before its parasitism of the eggs. In all cases, eggs in the early embryonic developmental stages prior to the gastrulation process are more vulnerable to infection. Once the hyphae is in contact with the egg a series of ultrastructural changes occurs in the egg shell due to the effect of exogenous metabolites and to the chitinolytic activities of the fungus.

Biocontrol Potential of *P. lilacinus*

The culture filtrate of the soil hypomyceterous fungus, *P. lilacinus* has been reported to have nematotoxic activity, efficient against many plant parasitic nematode species. The nematicidal action of culture filtrates is attributed to fungal metabolites or antibiotics produced by *P. lilacinus*. This fungus has been reported to produce peptidal antibiotics *viz.*, lilacinin, leucinostatin and Paecilotoxin. Acetic acid produced abundantly during fungal growth in liquid medium, inhibited movement of infective juvenile nematodes. The lytic enzymes like serine protease, chitinese and glucoamylase also play a role in penetration of the fungus through the egg shell of nematodes. *Paecilomyces lilacinus* has been used successfully to manage root knot nematode in tomato, banana, betelvine, carrot, groundnut and potato. It has also been reported to control *Tylenchulus semipenetrans* in citrus *Radopholus rabica* in banana and *Globodera rostochiensis* and *Pratylenchus* in potato.

Mechanisms of Action on Nematode

The fungus infects nematode eggs, produce an appressorium that adhere to the eggshell by mucilaginous material and develop infection peg to penetrate the eggshell. The size of an appressorium is dependent possibly on the chemical composition at the point of entry on the surface of the host.

Pasteuria penetrans an Obligate Biocontrol Agent

Pasteuria penetrans an obligate parasitic bacterium for the management of plant parasitic nematodes. It is a mycelial, gram positive endospore forming bacterial parasite suppressive to root-knot and few cyst forming nematodes. The parasite has been widely distributed in the world and associated with 323 nematode species under 116 genera from 69 countries in 5 continents (Table 2, 3).

Table 2: Effect of *Pasteuria* on nematodes

Nematode affected	Effect
Meloidogyne spp.	Larval mobility affected
	Reduction in nematode penetration
	inhibited egg formation
	Increased crop yield
Heterodera cajani	Reduced penetration
	Reduction in juvenile and cyst population in the soil and improved plant growth

Table 3: *Pasteuria* species as a potential biocontrol agent of root-knot and cyst forming nematodes

Nematode	Crop	Dose
M. incognita	Greengram	2×10^6 spores / m^2
H. cajani	Cowpea	1×10^7 spores / pot
M. incognita	Blackgram	4×10^6 spores / pot
M. incognita	Bhendi	2×10^6 spores / m^2
M. incognita	Tomato	2×10^6 spores / m^2

Biocontrol Potential

Interest in biological control of nematodes using *P. penetrans* gained momentum only after 1980, when Stirling and Wachtel discovered a technique for mass production of the bacterium. They observe that dried root powder containing spores of *P. penetrans*, when incorporated in soil @ 100 mg/kg soil resulted in 99 per cent infection of second - stage juveniles of *M. javanica* within 24 hours. Subsequently several researchers established the potentiality of *P. penetrans* as a biological control of nematodes.

3. Plant Growth Promoting Rhizobacteria as Biocontrol Agent

Bacteria that colonize the rhizosphere soil under the chemical influence of the root or that of rhizosphere competent are commonly referred to as rhizobacteria. Specific rhizobacteria have the ability to improve plant growth and/or root health. The term plant growth promoting rhizobacteria (PGPR) was used to distinguish these bacteria from other groups unable to promote plant growth.

Seven to ten per cent of the rhizobacteria isolated from the root system of potato, sugarbeet and tomato exhibited antagonistic activity towards different cyst and root-knot nematodes.

Bacillus

Bacillus thuringiensis suppressed root knot under field conditions (Prasad and Tilak, 1972). *In vitro* studies established that the exotoxin produced by the bacterium during growth and sporulation had high nematicidal activity against juveniles and eggs of *Meloidogyne* spp. Delta endotoxins of *B. thuringensis* var *isarlensis* and *B. thuringensis* var. *karstaki,* applied to soil showed decrease in egg laying and juvenile emergence from cysts of *H. glycines* (Sharma and Gomes, 1996).

Application of a nematicidal strain of *B. thuringensis* to tomato and pepper in Puerto Rico reduced galling in roots due to *M. incognita* and soil population of *Rotylenchulus reniformis* and increased yield significantly.

Bacillus subtilis was found to be antagonistic to *Meloidogyne* spp, *H. avenae, H. cajani* and *H. zeae*. This bacterium was able to suppress both *H. cajani* and *F. udum* in pigeon pea the combination of *Bacillus* sp, *Agrobacterium, Pseudomonas* and *Burkholderia cepacia* was found to be antagonistic to *M. incognita*. The root galling pattern on tomato was radically different in all rhizobacteria treated plants which showed minute round galls almost resembling *M. hapla* infection with a profuse root system (Jonathan *et al.,* 2000).

Bacillus cereus was reported to be antagonistic to *M. incognita* and *H. avenae, H. cajani,* and *H. zeae*. Exposure of *M. javanica* second stage juvenile to *B. cereus* in soil inhibited the penetration of the nematode into tomato roots. *Bacillus cereus* resulted in significant reduction in root galling and enhanced seedling biomass.

Agrobacterium

Agrobacterium radiobacter (strain G12) was found to reduce *G. pallida* juvenile penetration significantly in potato roots. The ability of the rhizobateria to stimulate induced systemic resistance against the nematode was demonstrated in split root system studies.

Burkholderia

The efficacy of plant growth promoting rhizobacteria, *B. cepacia* against *M. incognita* race 1 on tomato and banana was studied. The root galling pattern on tomato was radically different in the treatment which showed minute galls almost resembling *M. hapla* infection with profuse root system. The bacteria also limited reproduction of *M. incognita.*

The bacterium, *B. cepacia* (strain Bc-2) were investigated for activity against the nematode, *M. incognita*. It also suppressed eggs and J2 per g of root (9 % suppression) compared to water control.

4. Mechanism of Action of PGPR

Antibiotic-Mediated Suppression

It is well established that antibiotic production is one of the main mechanisms involved in the suppression of plant pathogens which includes both nematodes and pathogens by fluorescent pseudomonads. Antibiotics include phenazines, tropolone, pyrrolnitrin, pyocyanin and 2,4- diacetylphloroglucinol were produced by the bacterium.

Siderophore-Mediated Suppression

Pseudomonas fluorescens generally produce yellow green, water soluble, fluorescent siderophores with both hydroxamate and phenolate group. The production of these siderophores has been linked with the disease suppression ability of certain *Pseudomonas* spp. The efficient iron uptake mechanism also a significant factor for aggressivel colonization of plant roots thus aiding the physical displacement of deleterious organisms.

Induced Systemic Resistance

Induced Systemic Resistance (ISR) is based on plant defence mechanisms that are activated by inducing agents. ISR once expressed activates multiple potential defence mechanisms that include increased activity of chitinases, b-1,3-glucanases, peroxidases and other pathogenesis-related proteins; accumulation of antimicrobial low molecular weight substances e.g., phytoalexins and also formation of protective biopolymers, e.g. lignin, callose and hydroxyproline-rich glycoproteins.

5. Conclusion and Future Thrust

Though several biocontrol agents were discussed in this topic only very few *viz. P. penetrans P. fluoresens P. lilacinus, T. viridae* and AM fungi are found to be effective in field conditions against plant parasitic nematodes. Hence these bioagents can be included in the overall IPM strategies wherever it is feasible under different situations. In future there is a possibility of altering host specificity and improving virulence of biocontrol agents through biotechnological interventions.

References

Prasad, S.S.V. and K.V.B.R. Tilak, 1972. Aerobic spore forming bacteria from root-knot nematode infested soil. *Indian Journal of Microbiology* 11: 59-60.

Sharma, R.D and A.C. Gomes, 1996. Effect of *Bacillus spp* toxins on oviposition and juveniles hatching of *Heterodera glycines. Nematologica Brasileria* 20: 53-62.

Sivasithamparam, K and E.L. Ghisalberti. 1998. Secondary metabolism in *Trichoderma* and *Gliocladium* In: *Trichoderma* and *Gliocladium* vol.1 (Eds) C.P. Kublick and G.E. Harman. Taylor and Trancis, London: 139-191pp.

Jonathan, E.I.K. Barker, R., Abdel-Alim, F.F., Vrain, T.C. and D.W. Dickson, 2000. Biological control of *Meloidogyne incognita* on tomato and banana with rhizobacteria, actinomycetes and *Pasteuria penetrans*, *Nematropica*, 2: 231-240.

❐❐❐

29

Scope of Organic Amendments for the Management of Plant Parasitic Nematodes

JONATHAN, E.I.*, P. SENTHILKUMAR, K. DEVRAJAN AND S. PRABHU
Department of Nematology, Tamil Nadu Agricultural University, Coimbatore 641 003.
*Director, Centre for Plant Protection Studies, TNAU, Coimbatore-641 003.

1. Introduction

Plant parasitic nematodes threaten agricultural crops throughout the world, particularly in tropical and sub-tropical regions. The degree of damage to a particular crop is influenced by the crop and cultivar, nematode species, level of infestation and environment. Severe damage may result if high infestation levels occur in soil where susceptible crops are grown. These deleterious effects on plant growth result in low yield and poor quality. Crops such as banana, citrus, grapevine, coconut, coffee, cotton, peanut, pineapple, potato, rice, soybean, tobacco and vegetables are affected. Nematodes often cause decline or death of highly prized ornamentals and turf. In India crop losses in India due to nematodes is estimated at Rs. 24210 crore annually.

The symptoms of nematode damage on plants may be observed on shoots and/or roots. For example, root lesions, root pruning, root galling and cessation of root growth are common symptoms of nematode feeding. Roots damaged in this manner cannot efficiently use the moisture and nutrients available in the soil. Symptoms of nutrient deficiency, such as yellowing of foliage, result even when there is an adequate supply of nutrients in the root zone. Furthermore, roots so weakened and damaged are easy prey to many types of bacteria, fungi and other organisms which invade the roots and accelerate root decay.

For nematode management, there are two practices i.e. chemical and non-chemical methods. The chemical nematicides, although very effective but have certain limitations like high cost, environmental pollution, hazardous effect on human beings and animals. Under this situation, efforts should be made to concentrate on the use of non-chemical methods. Like cultural practices (crop rotation, treatment of bare root plants, steam sterilization of soil), This practice has received much attention by the researchers after a report that incorporation of chopped plant parts into root-knot infested soil reduced galling. This finding has produced an explosion of research on organic matter and their often irrational incorporation of anything, anyway, everywhere and in any amount into soil for nematode control. The beneficial effects of organic amendments is well demonstrated by various workers. It is an established fact now that organic amendments have the capacity of influencing and modifying the relationship of plant and nematodes. Due to this reason only major attention is being given to research on the prospects of organic amendments. Any soil amendment when incorporated in nematode infested fields, changes the soil environment and adversely affects the life cycle of nematodes and also enables the plant to resist their attack.

Various substances used as organic amendments include green manure, organic manures, dry crop residues, oilseed cakes, chitin and cellulose waste, meals, oil and sugars.

2. Mechanism of Action of Organic Amendments on Nematodes

Organic amendments of soil in reducing the damage caused by plant parasitic nematodes is now well established. However, the efficacy of organic amendments varies. It depends mainly on nature of the amendment, soil temperature and moisture, host type, nematode species and kind of soil biota. Reduction of nematode population may arise due to any of the following factors or may be due to combination of several factors.

2.1 Alteration of Physical and Chemical Conditions of Soil

Addition of organic amendments affects the physical and chemical conditions of soil. The increase in aeration and solubilization of nutrients results in better plant growth. The reduction in the population of citrus nematode, *Tylenchulus semipenetrans* was attributed to high nitrate levels and a decrease in soil pH reported decreased number of adult females in plant roots and postulated that this decrease could be due to lack of attraction of larvae to roots. Crop residue decomposition products are also known to influence physiology of the plant. It seems probable that in amended soil favourable conditions for rapid root growth, better and prolonged utilization of nutrients helps in reducing the nematode damage.

2.2 Parasitism and Predation of Plant Parasitic Nematodes by Nematophagous Fungi, Bacterial and Soil Microfauna

Role of nematode trapping fungi is one of the mechanisms of reduction of plant parasitic nematodes in amended soils. Decomposible amendments stimulate the growth of nematophagous fungi.

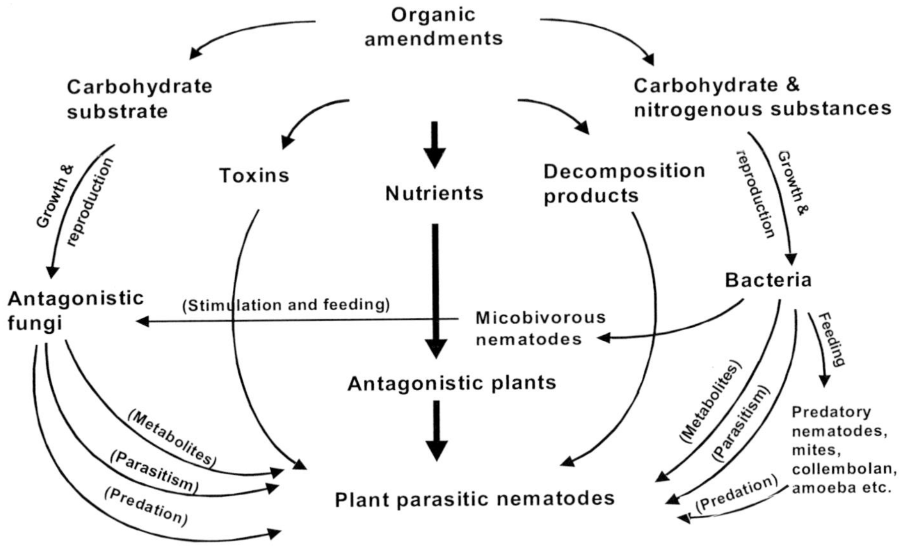

Fig.1. Mode of action of Organic amendments.

2.3 Production of Toxic Decomposition Products

A wide variety of chemically different substances are released during the microbial activity in amended soil or released as decomposition products during microbial breakdown of organic amendment which may be directly toxic to nematodes. Production of ammonia due to decomposition of waste and water hyacinth reduced the number of J_2 of *M. incognita*.

3. Green Manures

Green manures like *Sesbania aculata* and *Crotolaria juncea* have been found effective against plant parasitic nematodes. In betelvine crop. Reduction in phytonematodes associated with the crop by amending the beds with chopped and shade dried leaves of *Calotropis giantea* leaves has been reported. Significant reduction in the infestation of spiral nematode in sugarcane crop intercropped and incorporated with either marigold or daincha coupled with the application of either pressmud or neem cake has been observed.

Chopped plant parts (leaves, stem and flowers) of tobacco, cabbage, cauliflower, crotolaria, cassia, datura, sunhemp marigold, neem, bhang, subabool and periwinkle have shown good promises for nematode control.

Aqueous extracts of *Datura stramonium, Ricinus communis* and *Azadirachta indic*a significantly inhibited hatching of M. incognita and also increased the plant growth and reduced galls in okra. Singh and Singh (2002) reported that *T. minuta, C. sativa, P. rhoeas and A. barvedensis* were highly nematicidal and fresh extracts inhibited hatching of *M. incognita.*

Spot application of neem cake significantly increased plant growth and reduced the galls caused by *M. incognita* in French bean. Saravanapriya and Sivakumar (2004) reported that leaf extracts of *Calotropis gigantea* and latex of *Carica papaya* were effective in inhibiting hatching and caused highest percent mortality of *M. incognita* juveniles. Application of neem seed kernel extract, castor and karanj cake as seed dressing and soil application (2.5q/ha) improved the growth of cowpea and reduced *R. reniformis* infestation. Application of fresh leaves of *A. indica, Sesbania rabica* and water hyacinth compost @ 60kg N/ha effectively reduced *Hirshmanniella oryzae* population in rice and it also increased grain yield. Sobita Devi (2002) reported that application of *Ocimum sanctum, Annona squamosa, Eucalyptus* sp., *Tagetes* sp., *Ficus religiosa, Ricinus communis, A. indica, Myristica melabarica* and *Accacia rabica* significantly reduced the infection of *H. cajani* in mung, cowpea and urd bean.

4. Organic Manures

FYM and compost have been reported to suppress nematode population. According to Johnson, every type of dried crop residues / debris in soil reduced root-knot severity when residues are allowed to decompose for atleast 30 days (Ahmed and Choudhury, 2004). Sawdust and wheat straw decreased the cyst nematodes of *Heterodera avenae* on wheat. Reduction of number of galls produced by *M. incognita* in cotton in FYM, poultry manure, biogas slurry and spent compost in comparison to unamended soil has been reported. Reduction of root-knot index of *M. javanica* in bottlegourd and okra with poultry manure was also observed by recorded. Significant reduction in the infestation of root knot and spiral nematode in banana treated with pressmud @15t/ha and the treatment also enhanced the yield has been recorded (2004). Ahamed and Choudry reported that application of poultry manure @ 2t/ha were highly effective against *M. incognita* in French bean. Nisha and Sheela (2002) reported that application of green leaf mulch of neem chromolaena and glyricidia significantly increased crop growth and weight or rhizomes of medicinal plant *Kaempferia galangal* against root knot nematode.

Dry Crop Residues

It has been established that every type of dried crop residues/ debris in soil reduced the root-knot severity. Hay of alfalba, flax, timothy, Lucerne, straw of oat, rye, wheat, cotton waste, paddy husk have been tested successfully against plant parasitic nematodes.

5. Oilseed Cakes

Incorporation of castor bean pomace in soil/significant reduction of plant parasitic nematodes was reported. Both edible and non-edible oilcakes are used for suppressing nematode population in soil. Oilcakes of neem, groundnut, castor, linseed, mustard, mahua, cotton, sunflower, safflower, 'til' have been used for controlling root-knot nematodes in different crops. Soil amended with different oilcakes not only reduces the nematode infestation and root galling but also reduced the nematode population in soil and roots (Singh and Dabur, 2004).

In papaya application of krang cake @ 2.5t/ha at the time of pit preparation significantly increased the fruit yield and reduced the root knot infestation. Application of 10t/ha of FYM along with *T. viridae* as seed treatment @50g/kg of seed significantly reduced the root knot population and improved plant growth of soyabeen. Application of neem seed powder and *Paecilomyces lilacinus* @1 $\times 10^6$ spores/ml found to be effective against both reniform and root knot in brinjal. Borah and Phokan (2004) showed that compatibility of neem with *Glomus fasiculatum* against *M. incognita* in brinjal. Combined application improved the plant growth parameter and reduced the nematode population.

6. Chitin and Cellulose Wastes

Saw dust has effectively suppressed the various nematodes, due to their high C/N ratio.

7. Meals

Cornmeal, soybean meal, cotton seed, castorbean, bone meal and horn meal have been found very effective against root-knot and stunt nematodes.

8. Intercropping

Application of *Verticillium chlamydosporium* intercropped with marigold (1:6) recorded maximum plant growth and minimum galls and egg mass against root knot in tomato. Dhangar *et al* (2002) reported planting of *Tagetes* as intercrop in alternate row was highly effective in managing root knot in brinjal. Better plant growth and reduced root knot index was also observed. Intercropping of *Tagetes erecta* in banana significantly reduced *Pratylenchus coffeae* population and increased yield of banana up to 12.5kg/plant. Inter copping of sugarcane with *Brassica campestris* at two rows recorded highest reduction (23.7%) in nematode population. This sequence showed prolonged toxicity effectively against nematodes.

9. Future Perspective

There is absolutely no doubt that these amendments increase yield of crop and decrease the nematode population whatever the mechanisms may be involved. In future it could be the one of the best alternative to chemical for nematode management. Although organic amendments and phototherapeutic substances have been found effective in managing the nematodes, yet their application at field scales does not seem to be economical due to requirement of large amount of such materials.

References

Ahmed, J.A, and B.N.Choudhury, 2004. Management of root-knot nematode, *Meloidogne incognita* using organic amendments in frenchbean. *Indian Journal of Nematology*, 34(2): 137.

Borah, A and P.N. Phukan, 2004. Comparative efficacy of *Glomus fasciculatum* with neem cake and carbofuran for the management of *Meloidogyne incognita* on brinjal. *Indian Journal of Nematology*, 34(2): 129.

Dhangar, D.S. D.C. Gupta and R.K. Jain, 2002. Studies on intercropping of marigold (*Tagetes* sp.) with brinjal on plant growth and population of root-knot nematode (*Meloidogyne javanica), Indian Journal of Nematology*, 32(2): 220.

Nisha, M.S. and M.S. Sheela, 2002. Effect of green leaf mulching for the management of root-knot nematode in Kacholam, *Indian Journal of Nematology*, 32(2): 211.

Saravanapriya, B. and M. Sivakumar, 2004. Effect of some plant extracts on *Meloidogyne incognita* (Kofoid and White, 1919) Chitwood, 1949 juvenile mortality, *Indian Journal of Nematology*, 34(2): 180

Singh, K and K.R.Dabur, 2004. Effect of aqueous extracat of neem (*Azadirachta indica*) on egg hatching of *Meloidogyne incognita. Indian Journal of Nematology*, 34(2):133.

Singh, S.P. and V. Singh, 2002. Evaluation of Some Plant Extracts against Root-knot Nematode, *Meloidogyne incognita, Indian Journal of Nematology*, 32(2): 226.

Sobita Devi, L. 2002. Management of *Heterodera cajani* through the soil amendment of botanical leaves, *Indian Journal of Nematology*, 32(2): 206.

□□□

SECTION - IX

MISCELLANEOUS

30

Appropriate Plant Nutrition Technologies for Sustainable Pest Management

RAVICHANDRAN, M.
Professor and Head, Department of Soil Science and Agricultural Chemistry, Faculty of Agriculture, Annamalai University, Annamalainagar-608 202, Tamil Nadu

Global population increases annually by about 80 million and will hit the 8 billion mark within the next 20 years. More people need more food. In addition to that, with increasing income, quality becomes an important parameter in selecting the food . Consumer look more than before whether food is healthy and safe, and thus free of pests, but also free of residues of pesticides.

Most of the required higher crop production has to come from increased yield because land reserves for cropping become progressively scarce due to various reasons. However, growth in crop output is jeoparadised by losses caused from pests. To protect plants from pests and diseases, farmers have spent worldwide in 1998 a total of US $ 34 billion and will spend more with an annual growth rate in pesticide consumption of 4.4 %.

To control pests, the farmers have several options, which can be combined in the IPM approach.

- Genetics: *i.e.* the cultivation of crops, which are less susceptible or even resistant to pests

- *Biological control*: this refers to utilization of predators
- *Chemical control*: through organic / inorganic pesticides
- *Cultural practice* : to create optimal growth conditions of the cultivated crops and / or eradicate those conditions, which are favourable for multiplication of pests
- Plant nutrition

Fertilizers are backbone of food security as for the Indian Agricultural Scenario is concerned. They are absolutely indispensable for the overall health of agricultural production and productivity in this country.

Nutrition of plants has a substantial impact on the predisposition of plants to be attacked or affected by pests. By affecting the growth pattern, the anatomy and morphology and particularly the chemical composition, the nutrition of plants may contribute either to an increase or decrease of resistance and/or tolerance to pests. However, unlike in human nutrition where the effect of nutrition on "health" has gained considerable importance, the implementation of "healthy" nutrition to improve resistance and tolerance of plants lags behind its potential.

1. Mineral Nutrition and Resistance of Plants

According to Fuchs and Grossman (1972) adequate nutrition and fertilization can influence the resistance of plants to biotic stress in the following ways

- By increasing the density of stands, modifying the climate within the stand and even modifying the colour as a result of fertilizer application they can increase the infection stress.
- Caustic fertilizers such as slaked lime, liquid ammonia or urea and top dressing containing micronutrients may have a lethal effect on pests or reduce their virulence
- Accumulation of ions resulting from fertilizer use can damage pest also through osmotic mechanisms
- The different kinds of organic fertilizers (dung, compost, green fertilizers) can activate the organisms in the soil and thus encourage the development of antagonists to pests, thus raising antiphytopathogen potential of the soil.

- Well fertilized and adequately nourished plants are able to withstand stress better also on account of their higher regenerative capacity
- The use of fertilizers can also result in genuine resistance, however,either as a result of anatomical changes (such as thickening of cuticle, silification of tissue, tissue barriers, greater pilosity of the epidermis etc) or of changes in biochemical factors (synthesis of inhibitors, absence or deficiency of specific nutrients or metabolites)

2. Effects of Nutritional Status on Resistance of Plants

In view of the multitude of functions performed by mineral elements in plants they can influence plant resistance. There are far more than 1000 enzymes acting simultaneously or sequentially according to genetically fixed plan in only a single cell. Mineral elements involve in enzymatic reactions as elements or active groups of the enzymes themselves and by influencing and catalyzing enzyme synthesis and the metabolic reactions in which they take part.

Micronutrients in particular play an important role in the reaction mechanisms of defence enzymes. First as catalysts in the intermediary metabolism they influence or modify the composition of the protein, i.e. one half of defence enzymes, in living cells. Secondly, micronutrients are implicated in the synthesis of the prosthetic groups of defence enzymes, may even be an integral part of the prosthetic group, or atleast act in some way as catalysts, activators or inhibitors.

The situation is similar with macronutrients, particularly in the case of K and Mg, which control numerous enzyme reactions involved in carbohydrate, protein and nucleic acid metabolism. N and S are contained in proteins and are therefore essential to enzyme sysnthesis, and P play a decisive role as an important element in energy and nucleic acid metabolism.

The mineral elements can influence enzyme reactions by following ways:

- By forming prosthetic group of the enzyme, particularly in the case of micronutrients
- By forming active parts of the prosthetic group or being an integral part of the enzyme

- By their integrative function permitting substitution of one element for another in some cases
- By acting as facultative activators, and
- By inhibiting enzyme activity

It should always be remembered that in plants pest attack are possible only if the plant offers the pest concerned suitable living conditions, and this can be regulated through the metabolism of the plant. For instance the accumulation of certain products of metabolism such as amino acids or sugars can lead to stronger attack. It is therefore important to know how and in what direction metabolism in plants can be influenced by controlling the supply of mineral elements.

3. Influence of Nutrition as a Whole

Conclusions regarding the effect of nutritive conditions should always be considered critically because, it cannot be simply assumed that the optimal use of fertilizer in terms of economic yield is necessarily also the best from the phyto - medical standpoint. Moreover, in terms of performance, resistance in its restricted sense, tolerance and regenerative capacity can be mutually supplementary and compensatory. Moreover, resistance does not rise in the same direction with the supply of available nutrients, but may sometimes do exactly the opposite, particularly if the optimum for plant growth has been exceeded. For example apart from N, K has the greatest influence on resistance, and usually in the opposite direction to that of N, so that the level of resistance to pests depends on the N:K ratio. In general the influence of K on resistance predominates.

The effects of fertilizer application on insects feeding on crop plants also describe both the promotion and inhibition of growth. Although the incidence of bollworms (*Heliothis zea*) has been observed to increase following increased fertilizer application, mortality among the larvae of various forest pests increases. The nature of the mechanisms promoting or inhibiting pest development may therefore differ, but in all cases they are influenced in some way by metabolism. A clear picture, however will only be obtained by analyzing the effects of fertilizers on the absolute and relative concentration of mineral elements in plant tissues, the consequent changes in metabolism and its intermediary and end products, and the resulting morphological changes instead of merely recording how fertilization affects the incidence of pests.

It is necessary above all to find out the conditions under which the plant concerned achieves maximum protein synthesis. Resistance is higher in plants producing more protein than their conspecifies. The more fertile the soil, the lower is the incidence of pest attack, although " balanced fertility" in mind i.e without a surplus or deficiency of any element.

4. Effect of Nutrient Element on Pest

There can be no doubt that among the mineral elements N has the greatest influence on N resistance. In particular, excessive applications relative to other nutrients increase plant susceptibility almost without exception, and this fact must receive greater attention than hither to in connection with the increased use of mineral fertilizers. Changes in sugar content of the platns due to high rates of N fertilizers application are believed to be among the reason for differences in the severity of attacks. On the other hand, N fertilization which is particularly effective against insects feeding on plants or their organs, can ecncourage infestation with sucking insects unless large amounts of K are given at the sametime, because the K mineral has a specific inhibitory effect on sucking insects. The relative concentrations of N and K are more important than the absolute concentrations in the plant because both a relative N surplus and a relative K deficiency have roughly the same effect: a reduction in resistance.

Trichomes impart general antixenosis resistance by providing effective barrier that prevents small anthropods from landing on the plant surface and prevents movement and feeding (Southwood, 1986). This trichome density was found to decrease with increased level of N upto 180 kg ha^{-1}. Increase in N level increased the length and breadth of the leaves which favour the larvae to hold and feed the leaves.

It must also be mentioned that the different forms in which N is supplied to plants also affect plant resistance differently, depending on their physiologic reactions during metabolism. Urea for instance, in contrast to other forms of N fertilizer, is said often to increase plant resistance and is reputed to inactivate the toxins.

Phenolic compounds play an important role in the host/pathogen relationship being the basis for many defence mechanisms. They act as phytoalexins or as precursors of lignin and suberin which act as mechanical barrier. With unbalanced ample N supply, there is a high demand for carbon from photosynthesis via krebs cycle for soluble organic N compounds such as phenols and quinines. Under N limited condition, however, much more carbon from the krebs cycle is available for the synthesis of phenolic compounds.

Excessive growth due to unbalanced N supply can also create microclimate conditions favourable for pests. Lodging of cereals as commonly observed at over-supply with N and inadequate K is a good example, humidity remains longer in lodged crops giving ideal conditions for pests.

It is assumed that plant resistance can be improved by maximizing protein synthesis ("pure protein" is meant here), it is readily understandable that N can reduce resistance in one case and increase it in another. The application of N fertilizers will negatively affect the resistance of a plant only if N compounds with a low molecular weight accumulates during the excessive amounts of fertilizers or protein synthesis is inhibited owing to a deficiency of other elements.

Research has shown that crops with high N content enable the pests to produce more eggs, survive better, live longer and become ecologically fitter (Heong, 2005). Increase in doses of N enhanced gall midge stem borers, leaffolders and leafhoppers, BPH; leaffolder; striped stem borer-*Chilo suppressalis Wlk* and green leafhopper, BPH, rice bugs, leaffolders and stem borers in rice; *Bemesia tabaci Genn.* In cotton; leafhopper and aphids leafhopper and white fly in brinjal; *Aphis gossypii* on potato (Singh *et al.*, 2005).

Potassium is involved in numerous functions in the plant such as in enzyme activation, cation/anion balance, stomatal movement, phloem loading, assimilate translocation and turger regulation etc. stomatal resistance decreases and photosynthesis increases with increasing K content of leaves (Peoples and Koch). K in contrast to N is said to have mainly positive effects on resistance. Normally, a plant well supplied with K incorporate more protein from N assimilated. Potassium deficient leaf cells accumulate substantial quantities of low molecular weight organic compounds because they act as an osmoticum in the absence of sufficient K. Leaf sheaths of rice given only NP but no K contained nearly three times the amount of soluble N and more than twice the content of soluble sugar than plants fertilized with adequate K. Asparagines putrescine and agmatin etc are reported to accumulate if the N: K balance is disturbed, and this in turn makes the tissue physiologically more prone to infection; soluble amides and amino acids are also easier for the phytopathogens to utilize. In addition K deficiency leads to thinner cell walls and slower growth of meristematic tissue. This makes it easier for sucking insects to penetrate the epidermis. The reduced insect attack is a result of stronger cell walls, the more optimal form of leaf hairs that deter insects and the lower rates of sugar synthesis.

The reduction of red spider mites (*Tetranychus spp.*) on tomatoes is reportedly encouraged by K fertilization, whereas in chrysanthemum K deficiency has this effect. On the other hand, K fertilization reportedly reduce the colonization of linseed plants by thrips (*Thrips lini*) and of various other plants by aphids. Aphid infestation of maize can increase by a factor of four as a result of K deficiency, and the larvae of the cereal leaf beetle fed only on leaves with a low K content. Proliferation of the bean aphid in contrast, is said to increase as rates of K fertilization increase, and the reproduction of green peach aphids is inhibited by K deficiency. Highest resistance to BPH could be provided if the N:K ratio in fertilizer was 1:1.5. In case of some other insect pests, like thrips, green hoppers, whorl maggot and leaf roller, similar affects are reported.

K fertilizer that contain chloride are occasionally reported to reduce resistance in plants but chloride is superior to those containing sulphate in respect of resistance enhancement (Perrenoud, 1990). Insufficient K also causes a pale leaf colour that is particularly attractive to aphids .Wilting commonly observed with K deficiency, is another attraction to insects. Cracks, fissures and lesions that develop at K deficiency on the surface of leaves and fruits provide easy access to insects. IPI sponsored on-farm trials in India with soybean showed considerable less incidences with gridle beetle, semilooper and aphids when applied with adequate potash.

Phosphate like K is said to enhance the formation and stability of tissues. High rates of P fertilization are said to reduce the damage caused by true bugs and leaf eating pests to Lucerne and by caereal leaf beetles. P fertilizer that were rich in silicates increase the resistance of plants to pests. Calcium is indispensable for the integrity and stability of cell walls. With insufficient Ca, cell walls leak low molecular weight organic compounds used as feed by certain insects. Since Ca strengthens the cell wall makes them more difficult for sucking insects to penetrate. Liming is also said to inhibit reproduction in aphids, cereal bugs (*Toxoptera graminum*) and red spider mites (*Tetranychus sp.*) .Conversely, increased aphid attack on Lucerne after liming has also been reported.Magnesium, on account of its influence on energy, protein and particularly nucleic acid metabolism, can be expected to greatly simulate cellular defence reactions. Increased colonization of tomato plants by red spider mites (*Tetranychus spp.*) of field beans by the bean aphid and of citrus by scales (*Lepidosaphes beckii*) have been reported to follow high rates of Mg fertilization.

Silicon is not considered as an essential element for most plants, but it has been reported that absorption of silicon is beneficial to crops, that induced protection against pest attacks (Epstein, 1994). The protection conferred to plants by silicon could be related to its accumulation and polymerization in the cells, forming a mechanical barrier that defies the insect pest attack (Yoshida *et al.*, 1962). A similar mechanism might contribute to plant resistance to insects such as stem borers (Ukwunguru and Odebiyi, 1985). Silicon fertilized plants present changes in their trichomes (Samules *et al.*, 1993), as well as in their epidermis and internal tissues (Hondson and Sangster 1988) defenses in wheat to green bug (*Schizaphis graminicum*) Flavia Batista Gomes *et al.* 2005). Factors, which lower soil pH lead to higher silicon uptake and content in plants . A positive effect of K on silicon uptake of rice was reported. Heavy application of N are known to make rice plants more susceptible to pest attack because of decreases in silicon concentration in the straw. Hence, application of silicon is recommended when high rates of N is followed. (Normally gramineae contain 10 – 20 times high silicon than legumes and other dicotyledons. Lowland rice commonly contain between 4.6 and 7.0 % silicon in straw).

5. NPK Fertilizers

In tomato, increase in the rate of NPK fertilizers reduced the major insect resistant traits namely trichome density as well as the amount of glycoalkaloid, 2-tridecanone (Barbour *et al.*, 1991).In contrast to synthetic fertilizers, organic manures often enhance insect resistance (Karungi *et al.*, 2006).Similarly, foliar spray of some micronutrients such as Fe, Mn and Zn exerts negative effect on insect pests (Mourad *et al.*, 2004).

6. Induction of Resistance by Bioinoculants

Mycorrhizae are symbiotic associations between plant roots and certain soil fungi which play a key role in nutrient cycling in the ecosystem and also protect plants against environmental and cultural stress. Arbuscular mycorrhizal fungi (AMF), which form symbiotic associations with root systems of most agricultural, horticultural and hardwood crop species, have been suggested as widespread potential bioprotective agents. AMF being biofertilizers and bioprotectors, an appropriate management of this symbiosis will enable considerable reduction of chemical fertilizer and pesticide inputs, especially in horticultural crops (Azcón-Aguilar and Barea, 1997). Fungal symbiosis in various crop plants confers tolerance and / or resistance against many phytophagous insects, nematodes and pathogens.Co-inoculation of AM fungus, *Glomus* spp. With *Azospirillum brasilense* (Linn.) reduced the incidence of leaf folder, *Cnaphalocrocis medinalis* Guen. In upland paddy var. PKM-1 (Amutha *et al.*, 2003) while individual inoculation of *G. intraradices* in pea plants conferred resistance against the adult weevils, *Sitona lineatus* (Wamberg *et al.*, 2003).

7. Conclusion

Naturally, it would be totally unrealistic to expect a simple link to exist between the nutritional status of plants or fertilizer use and the incidence of pests. Even the relationship between parasite and host plants is of an extremely complex nature, and the additional consideration of nutrients and fertilizers naturally makes it much more complex not simpler.

When using fertilizers, the primary aim must no longer be only to achieve higher yields, but to care at the same time for crop health, because this along with tillage and other aerable farming operations, represents an important factor affecting efficiency and the high yields of the desired quality will be achieved. The grower should not ignore the fact that consumers of his produce ask progressively more for "healthy" and "safe" food without residues of agrochemicals. The consumer prefers to purchase food that is "naturally" produced.

Balanced fertilization complies with this demand. Plants which are supplied with all necessary nutrients in a balanced manner are, as shown, more resistant to pests. This lower the need for particular pest control measures and reduces the risk of unwanted residues of pesticides. In terms of ecology and environmental protection, a balanced supply of mineral nutrients to the plants can also help reduce cost and the potential risk of environmental pollution by pesticides. With adopting balanced fertilization, the farmer can prove that he produced the food in a "natural" way and at the same time he is more competitive at the market.

References

Amutha, M., B.R. Mahadevan, K. Chozhan and G. Gajendran, 2003. Synergist effect of *Azospirillum brasilense* (Linn.) and vesicular arbuscular mycorrhiza (*Glomus* spp.) on management of leaf folder in rice. *Proc. Natl. Symposium on Sustainable Insect Pest Management*, 6-7 February, Entomology Research Institute, Loyola College, Chennai, abstr. P. 48.

Azcon-Aguilar, C. and Barea, J.M., 1997. Appliying mycorrhiza biotechnology to horticulture: significance and potentials. *Scientia Horticulturae*, 68 (1-4): 1-24.

Barbour, J.D., Jr. R.R. Farrar, and G.G. Kennedy, 1991. Interaction of fertilizer regime with host plant resistance in tomato. *Entomol. Exp. Appl.*, 60: 289-300.

Chellliah, S. 1986. Insecticide induced resurgence of rice brown plant hopper, *Nilaparvata lugens (Stal)*. In: Resurgence of sucking pests. *Proc. Natl. Symp.*, Ed., Jayaraj. P.1

Epstein, E . 1994. The anomaly of silicon in pant biology. *PNAS* (USA), 91: 11-17

Flavia Batista Gomes, Jair Compos de Moraes, Custodio Donizete dos Santos and Maarcio Marcos Goussain . 2005. Resistance induction in wheat pants by silicon and aphids . *Sci. Agric* .(Piracicaba, Braz.), 62 (6): 547 – 551.

Fuchs, W.H., and F. Grossmann, 1972. Ernahrung and resistenz von kulturpflanzen gegenuber Krank beitserregern and schadlingen. In: Hands, Pflanzenern, U. Dungung. Bd.1, Zweite Halfte, Springer, Wien – New York.

Heong, K.L. 2005. Strattegies to attain ecologically sustainable pest management in rice systems. In: Advances in Rice Science, Proceedings of International Rice Science Conference, Korea, (eds. K.S. Lee, K.K. Jena and K.L. Heong), pp.111-123.

Hondson, M.J. and A.G. Sanster, 1988. Observation on the distribution of mineral elements in the leaf of wheat (*Triticum aestivum* L.) with particular reference to silicon.*Annals of Botany*, 62: 463 – 471.

Karungi, J., B. Ekbom and S. Kyamanywa, 2006. Effects of organic versus conventional fertilizers and insect pests, natural enemies and yield of *Phaseolus vulgaris*. *Agri. Ecosys. Environ.*, 115(1-4): 51-55.

Mourad, A.K., S.A. Hammad, G.Z. Guirguis, O.A. Zaghloul and H.A. Sadek, 2004. Action of some micronutrients on the infestation and yield components of faba bean by the aphid, *Aphis craccivora* (Koch) (Aphididae: Homoptera) and the leaf miner, *Liriomyza trifolii* (Burgess) (Agromyzidae: Diptera). *Comm. Appl. Biol. Sci., Ghent Univ.*, 69(3): 291-304.

Perrenoud, S. 1990.Potassium and plant health. IPI Research Topics No. 3,2nd rev. edition .Basel / Switzerland.

Samuels, A.L., A.D.M. Glass, D.L.Ehret, J.G.Menzies, 1993. The effects of silicon supplementation on cucumber fruit: changes in surface characteristics. *Annals of Botany*, 72: 433- 440.

Singh K.L., L.Nabachandra Sing and M.P.Sindh. 2005. Influence of five nutrient levels with or without Azotobacter on the incidence of *Apis gossypii* (Glover) in potato. *Indian J. Ent.*, 67(2) : 165 – 169.

Ukwunguru, M.N. and J.A. Odebiyi, 1985. Incidence of *Chilo zacconius* Bleszynski on some rice varieties in relation to plant characters. *Insect Sci. Appl.* 6: 653- 656.

Wamberg, C., Christensen, S. and I. Jakobsen, 2003. Interaction between foliar-feeding insects, mycorrhizal fungi, and rhizosphere protozoa on pea plants. *Pedobiologia*, 47(3): 281-287.

Yoshida, S., Y. Ohnishi and K. Kitagishi. 1962. Histochemistry of silicon in rice plant. *Soil Sci. Nutr.*, 8: 107-111.

❑❑❑

31

Plant - Pest Interactions

PANNEERSELVAM, R.
Professor and Head, Department of Botany, Annamalai University,
Annamalainagar – 608 002, Tamilnadu

1. Introduction

Upon recognition of a pathogen, plants exhibit remarkable ability to differentiate pathogen, and later based on the nature of pathogen mount a resistance response meant to cease pathogen growth and disease development (Elzbieta Kuzniak and Maria Sktodowska, 2004). The resistance responses of plants include activation of local and systemic defenses (e.g. expression of pathogenesis-related proteins) and induction of a localized plant cell death at the site of infection called the hypersensitive response (HR). The HR is a form of programmed cell death (PCD) and shares some molecular and biochemical similarities with animal apoptosis. It is believed that the coordinated activation of PCD and defense mechanisms at the site of pathogen entry provides the plant with an efficient defense response that prevents pathogen proliferation and its possible consequence: systemic infection.

2. Saviors of Plant Defense

Salicylic acid (SA), nitric oxide (NO) and reactive oxygen species (ROS) (particularly H_2O_2) increase in abundance following pathogen recognition and each are important signaling molecules that promote and coordinate defense and HR responses (Magali, M. and C.J Bede, 2004). There exists complex synergistic (and possibly antagonistic) interactions between SA, NO and ROS that define the responses to biotic stress (Delledonne, 2005). The increase in ROS (the so-called oxidative burst) involves activation of a plasma membrane-localized nicotinamide adenine dinucleotide phosphate (NADPH) oxidase. During the HR, this is accompanied by an active down-regulation of ROS-scavenging systems to further promote ROS accumulation (Omprakash Mittapalli *et al.*, 2007). It is hypothesized that plant mitochondria act in the perception of biotic stress and take part in initiating responses such as the HR (Amirsadeghi, S. *et al.*, 2007).

3. Plants and Insects Interaction

The key mechanism used by plants to recognize herbivorous caterpillars is elicitors present in their oral secretions; however, these elicitors not only cause the induction of plant defenses but recent evidence suggests that they may also suppress plant responses. The absence of "expected changes" in induced defense responses of insect-infested plants has been attributed to hydrogen peroxide produced by caterpillar salivary glucose oxidase (GOX). Activity of this enzyme is variable among caterpillar species; it was detected in two generalist caterpillars, the beet armyworm (*Spodoptera exigua*) and the bertha armyworm (*Mamestra configurata*), but not in other generalist or specialist caterpillar species tested. In the beet armyworm, GOX activity fluctuated over larval development with high activity associated with the salivary glands of fourth instars. This implies that a factor in the diet is involved in the regulation of caterpillar salivary enzyme activity. Therefore, plant diet may be regulating caterpillar oral elicitors that are involved in the regulation of plant defense responses.

In response to herbivores, volatile terpenoid compounds are synthesized and released by many plant species. The synthesis of terpenoids affected by the herbivores at molecular level to control the expression of genes involved in metabolic pathway of terpenoid synthesis.

4. Plants Topical Self Suicide – A Mechanism to Protect

Localized activation of programmed cell death (PCD) in response to microbial attack is thought to act as a defense mechanism that inhibits the growth of pathogens within infected plant tissues. By killing cells at and around the site of infection this process generates a physical barrier composed of dead plant cells and limits the availability of nutrients to the pathogen because of rapid dehydration that accompanies tissue death, which subsequently accompanied by the induction of numerous antimicrobial defenses. Among these are pathogenesis-related (PR) proteins, such as glucanases and chitinases, and phytoalexins. PCD that occurs during the HR is accompanied by an increase in the production of reactive oxygen intermediates (ROI). Recent studies indicated that ROI in the form of H_2O_2 and O^ may be key mediators of PCD during the HR. ROI also were implicated as signal transduction agents that lead to the induction of other defense mechanisms such as PR proteins, salicylic acid (SA) biosynthesis (13), and systemic acquired resistance H_2O_2 was shown to induce PCD and defense mechanisms in bean and tobacco.

It is presumed that the recognition of an invading pathogen results in the activation of a plasma membrane-associated NAD(P)H oxidase and the excess production of O_2. Subsequent spontaneous or superoxide dismutase (SOD) catalyzed dismutation of O_2 leads to the accumulation of H_2O_2 that diffuses into cells and causes the activation of defense mechanisms and PCD. In accordance with this model, ROI scavenging mechanisms such as ascorbate peroxidase (APX) and catalase (CAT) were shown to be suppressed during the HR, a process that was suggested to potentiate the induction of PCD and other defenses because of the increase in ROI levels that results from the decreased capability of cells to scavenge H_2O_2 during the HR. Mitochondrial electron transport is associated with the generation of ROS such as super oxide and H_2O_2. Because ROS can damage macromolecules, their cellular levels are managed through avoidance and scavenging mechanisms (Mittler *et al.*, 1999).

5. Destiny of Mitochondria and Chloroplast

The signaling roles for SA during biotic stress, suggested its direct impact in mitochondria. It was shown that SA disrupts mitochondrial function in a concentration-dependent manner in tobacco suspension cells. At low concentrations, it acted as an uncoupler, whereas at higher concentrations it strongly inhibited electron flow. These effects were seen in both whole cells and isolated mitochondria and provide a rationale for studies showing that SA could dramatically inhibit ATP synthesis by tobacco cells (Amirsadeghi, S. *et al.*, 2007) also found that AOX expression correlated with the ability of SA to disrupt mitochondrial function. Hence, AOX may represent an excellent 'reporter gene' to evaluate whether mitochondrial dysfunction is occurring during biotic stress. Another signal molecule during biotic stress is NO, which along with SA and ROS, has been shown to promote the HR Plant cyt oxidase is similarly sensitive to NO but whether the physiological

NO concentrations generated during plant-pathogen interactions are sufficient to inhibit cyt oxidase and whether such inhibition contributes to defense responses or the HR remains unknown. Chloroplasts are a large source of ROS and so it is often assumed that the steady-state level of cellular ROS as well as the capacity of cellular ROS-scavenging systems is largely determined by this organelle.

6. Action of Host Versus Pathogen

An important set of virulence factors in pathogenic fungi is the so-called host-selective toxins that interact with host molecules to cause plant cell death and contribute to disease development. One such toxin, victorin, was shown to bind to and inhibit mitochondrial glycine decarboxylase (GDC), suggesting that GDC inhibition acted to promote cell death. Victorin treatment of oat leaves resulted in a loss of DCM, followed by an ability of victorin to gain access to the mitochondrial matrix. The cell death precedes access of victorin to the cell interior and that victorin likely interacts with a cell surface protein to initiate defense responses and cell death (Rentel C. *et al.*, 2008). In this respect, the virulence of the toxin may reside in its ability to elicit a plant PCD pathway. These results shed doubt on the importance of victorin-induced GDC inhibition in promoting cell death, but they do not preclude a role for the mitochondrion in this cell death. Victorin induces a burst of mROS preceding death. Harpins are virulence factors produced by bacterial pathogens such as *Pseudomonas syringae.* At low concentrations, H_2O_2 acts as a signaling molecule in the plant. During incompatible plant-pathogen interactions, endogenous H_2O_2 levels increase and have been implicated in localized cell death during the hypersensitive response of some plant species against invading pathogens. It was shown that a large and early ROS burst associated with harpin treatment emanated specifically from the mitochondrion, suggesting the ETC as the likely source of ROS (Mittler, R. *et al.*, 1999)

There is also evidence that the ROS-scavenging capacity of the mitochondrion is modulated in response to pathogen infection. In particular, increases in mitochondrial superoxide-scavenging capacity combined with decreases in the H_2O_2-scavenging components of the organelle were seen during *Botrytis cinerea* infection of tomato leaves and it was hypothesized that this could promote accumulation of mitochondrial H_2O_2 (Amirsadeghi, S. *et al*, 2007). Such results imply an active mechanism to ensure accumulation of specific ROS species at the mitochondrion.

Polyamines such as spermine are proposed to play a role during biotic stress responses. Spermine accumulates dramatically and in an N-gene-specific manner in the apoplast of tobacco mosaic virus (TMV)-infected tobacco because of up regulation of genes involved in spermine biosynthesis (Rentel, C. *et al.*, 2008). Accumulated polyamines are subsequently degraded in the apoplast by polyamine oxidase, generating H_2O_2 that may contribute to plant responses (Yoda *et al.*, 2003). Exogenous application of spermine to tobacco leaves could induce defense responses and cell death, mediated by a pathway involving activation of mitogen activated protein (MAP) kinases and resulting in increased expression of HR marker genes and transcription factors (Mit). It was also shown that AOX induction and activation of the MAP kinases could be blocked by the antioxidant flavone and by the Ca^{21} channel blocker La^{21} suggesting that increases in ROS and influx of Ca^{21} are cellular events.

The treatment of tobacco cells with A resulted in the rapid expression of eight different genes, as identified using differential display. Only one of these genes encoded an ETC component (AOX), whereas the others encoded proteins more generally implicated in either senescence or (biotic) stress responses. All eight genes were also induced by treatment of cells with SA or H_2O_2. Each of the gene-inducing treatments was associated with increased cellular levels of ROS and gene induction could be partially blocked by antioxidants that lowered ROS levels, suggesting ROS as an important intermediary in gene induction. The authors also showed that pre-treatment of cells with BA (Bonkrekic acid) blocked induction of all eight genes regardless of whether AA (Antimycin A), SA or H_2O_2 was used as the inducing agent (Maxwell *et al.*, 2002).

7. Conclusion

The potential regulatory role of ROI involved in pathogen infection leads to HR. During HR the ROI play in different roles to trigger the defense mechanism of plants to protect from the destruction nature of diseases. Many findings provide direct evidence for the involvement of ROI in mediating pathogen induced PCD.

References

Elzbieta Kuzniak and Maria Sktodowska, The effect of *Botrytis cinerea* infection on the antioxidant profile of mitochondria from tomato leaves, *Journal of Experimental Botany*, Vol. 55, No. 397, pp. 605-612, March 2004 .

Magali Merkx-Jacques and C. Jacqueline Bede, Caterpillar salivary enzymes: "eliciting" a response, *Phytoprotection* 85 : 33-37, 2004.

Maike C. Rentel, Lauriebeth Leonelli, Douglas Dahlbeck, Bingyu Zhao, and J. Brian Staskawicz, Recognition of the Hyaloperonospora parasitica effector ATR13 triggers resistance against oomycete, bacterial, and viral pathogens, *PNAS*, Vol. 105, No. 3, 1091–1096, 2008.

Omprakash Mittapalli, J. Jonathan Neal and H. Richard Shukle, Antioxidant defense response in a galling insect, *PNAS*, vol.104, No.6, 1889-1894, 2007.

Ron Mittler, Elza Hallak Herr, Bjorn Larus Orvar, Wim Van Camp, Hilde Willekens, Dirk lnze, and Brian E. Ellis Transgenic tobacco plants with reduced capability to detoxify reactive oxygen intermediates are hyperresponsive to pathogen infection, *PNAS*, Vol.96, no. 24, 1999.

Sasan Amirsadeghi, Christine A. Robson and C. Greg Vanlerberghe. The role of the mitochondrion in plant responses to biotic stress, *Physiologia Plantarum* 129: 253-266. 2007.

❑❑❑

Index

A

Abrasive dusts 118

Acceptable pest levels 106

Aceria guerreronis 89, 98, 100, 101

Achook 58, 246, 252

Activated gene 212

Alkaloids 18, 227, 228, 229, 240, 241, 242, 249

Antibiosis 6, 297, 298, 299, 301, 317

Antibiotic 167, 191, 193, 322

Ants 22, 130, 150

Ants 40, 41, 42, 46, 61

Aphidophagous coccinellids 150

Aphis craccivora 173, 174, 342

Apoptotic suppressor gene 215

Aspergillus terreus 223

Aspidiotus destructor 97

Automatic insect removal bin 268, 270, 271, 273

Azadirachtin 18, 66, 93, 95, 96, 100, 155, 252

B

Bacillus thuringiensis 26, 27, 28, 29, 35, 57, 62, 114, 125, 143, 198, 209, 211, 221, 320

Back break trap 118, 289

Bacterial antagonists 193

Baculoviruses 19, 33, 34, 154, 203, 204, 205, 206, 207, 208, 209, 210, 213, 215, 217

Bait stations 289, 290, 291

Bait trap 121

Balanced fertilization 341

Beauveria bassiana 16, 19, 20, 28, 31, 32, 33, 35, 73, 156, 188, 201, 221, 222, 224, 246

Biocontrol based IPM 183
Biointensive IPM 108, 109, 122
Biorational pesticides 125
Biota 55, 326
Biotech crops 114
Bird perches 52
Black headed caterpillar 90, 92, 95
Botanical pesticides 21, 57, 126, 229, 234, 235, 236, 254
Botanicals 18, 57, 58, 59, 63, 91, 93, 99, 100, 126, 142, 228, 232, 235, 236, 237, 250
Burkholderia 201, 321

C

Cardamom itch 76
Cardamom whitefly 77, 87
Centrifugation 36, 154, 155, 156
Chemical mutagens 210
Chitinase gene 214
Chrysanthemum 143, 229, 240, 250, 339
Chrysanthemum cinerariaefolium 229, 240
Community farming 7
Community level practices 115
Cutworms 45, 117, 275, 281
Cylas formicarius 44, 263
Cysteine protease 214

D

Delta trap 265
Dimboa 110, 242
Diuretic hormone gene 213

E

Edaphic factors 244
Electrocution traps 290
Emergence trap 121
Enhancin gene 213
Entomogenous fungi 20, 188
Entomopathogenic fungi 27, 30, 33, 159, 188, 246
Entomophthorales 32, 144
Entomopox virus 208, 209, 212, 218
Eriophyid mite 89, 90, 91, 92, 93, 98, 101
Eugenol 15, 54, 230, 288

F

Fish meal trap 287, 291
Flea beetles 44, 116
Flyash 305, 306, 307, 308, 309, 310, 311, 312, 313, 314
Flyash as a carrier 311, 312
Flyash herbal pesticides 311, 312
Flyash mycoinsecticides 312
Fungal antagonist 17
Fungal onslaught 223
Funnel trap 285

G

Giant african snail 39
Glucosinolates 241, 242
Glue trap 261, 289
Green food 12

H

Heirloom 14
Hopper dozer 118

I

Ifoam 3, 10, 25, 26, 107
Incandescent lamps 280
Insect probe trap 268, 291
Intercropping 13, 27, 53, 112, 115, 330, 331

IPM 9, 15, 20, 21, 22, 23, 26, 50, 52, 62, 63, 79, 90, 91, 103, 104, 105, 106, 107, 108, 109, 116, 122, 125, 126, 127, 142, 177, 180, 181, 182, 183, 184, 259, 260, 261, 322, 333
IPM practices 21, 50, 79, 126, 180, 184
ITK 20, 21

J

Jaivik agriculture 8
JHE 213

K

K deficiency 337, 339
K fertilizer 339
Kairomones 17, 66, 237, 239
Kaolin 125

L

Lures 119, 261, 263, 275, 276, 277, 278, 285, 286

M

Malaise traps 293
Mass trapping 258, 261, 262, 264, 268, 269, 272, 275, 277, 279, 286, 292, 294
Mating disruption 15, 258, 263
Mealy bug 59, 98, 115
Mechanical exclusion 119
Mechanical force 118
Melia azedarach 231, 232
Microbial pesticides 19, 21, 23, 28, 35, 57, 197, 198, 199, 200, 201, 219, 220, 221, 222, 225, 235
Mineral nutrition 334
Mitochondria 346, 347, 350
Mothcidal fungi 224, 225
Mulches 41, 113, 116
Mycoinsecticides 31, 156, 219, 220, 225, 226, 312, 314
Mycomite 71

N

Ncipm 20
Neem 12, 18, 21, 22, 45, 52, 54, 55, 58, 59, 60, 61, 63, 66, 85, 86, 92, 93, 94, 97, 98, 99, 100, 118, 126, 155, 183, 231, 232, 234, 235, 236, 237, 243, 244, 245, 246, 247, 250, 252, 253, 254, 255, 291, 309, 311, 312, 328, 329, 330, 331
Neem kernel aqueous extract 66
Nematodes 3, 35, 38, 75, 78, 79, 80, 86, 87, 101, 112, 113, 114, 123, 157, 158, 166, 198, 220, 221, 252, 315, 316, 317, 318, 319, 320, 322, 325, 326, 327, 328, 329, 330, 331, 341
Net barrier 51
Nicotiana tabacum 228, 229, 250
NPK fertilizers 340
NPV 16, 17, 19, 22, 27, 28, 35, 36, 84, 169, 170, 171, 172, 195, 196, 201, 204, 205, 206, 207, 208, 209, 211, 212, 213, 214, 215, 221

O

Organic amendments 326, 327, 330, 331
Organic farming 1, 2, 3, 5, 8, 9, 10, 12, 13, 20, 25, 26, 27, 28, 29, 34, 47, 62, 66, 89, 90, 91, 105, 106, 107, 161
Organic fertilizers 334

P

Paecilomyces farinosus 33, 144, 156
Panchamahabuthas 50
Pandora delphacis 33, 222, 223, 224, 225, 226
Parapheromones 276
Parasporal body 29

Phenolic compounds 18, 337, 338
Phenolics 242
Pheromone traps 21, 25, 45, 92, 109, 120, 258, 285
Physical control 82, 116
Phytohaemagglutinins 241, 242
Phytonematodes 166, 317, 328
Phytophages 178
PIB 35, 208
Pit fall trap 292, 294
Poison bait traps 286
Polyamines 349
Potassium 54, 62, 338, 343
Probe trap 268, 270, 273, 291, 292
Programmed cell death 345, 346
Protease inhibitors 242
Protinex mixture 148
Prudence 5, 7
Pure protein 338

R

Red palm weevil 276, 278, 286
Rhinoceros beetle 33, 89, 90, 92, 93, 94, 101, 118, 276, 286, 288
Robinson trap 120, 281, 282
Root grub 75, 77, 80, 81, 82, 84, 85
Rotenone 18, 229, 236, 237, 241

S

Sabadilla 229, 236
Saponins 241, 242
Secondary metabolites 18, 227, 239, 241, 243, 249, 298
Secondary plant compounds 239
Sevagram declaration 12
Shade regulation 80
Shot-hole borer 72
Silica 118, 125, 306, 307
Slug caterpillar 91, 96
Snap traps 290, 291
Spiders 16, 22, 55, 254, 294, 310, 311
Sprinkler irrigation 52, 53, 62
Sticky traps 52, 67, 82, 109, 117, 279, 284, 294
Strip cropping 12, 112
Suction traps 277
Sugar esters 125, 126, 230
Synergists 235

T

Tannins 18, 241, 242, 249
Tea mosquito bug 65, 71, 72
Tea thrips 67
Termites 13, 40, 41, 42, 45, 189
Terpenoids 18, 37, 227, 228, 241, 243, 249, 346
Thanjavur bow trap 289
Toxic amino acids 242
Traditional botanicals 228
Traditional methods 115
Traditional practices 61, 91
Trap baits 54
Trap cropping 13, 80, 115, 301
Trap crops 38, 51, 53, 54, 114
Tribal rat trap 290, 291, 294
Trichomes 299, 300, 302, 303, 337, 340
Turnaca acuta 96, 101

U

Uni trap 285

V

Viral unit 155
Visible radiation 117

W

Water trap 261, 265
Weevils 44, 60, 95, 264, 285, 341
White grub 14, 19, 89

Z

Zoophthora radicans 223, 224, 225

□□□

ABOUT THE EDITOR

Dr. P. Narayanasamy: Presently serving as Dean, Faculty of Agriculture and Professor and Head, Department of Entomology, Faculty of Agriculture, Annamalai University, Annamalai Nagar, Tamilnadu and has rendered 36 years of service at Annamalai University in the Department of Entomology. He conducted scores of research projects funded by Tamil Nadu State Council for Science and Technology, Government of Tamil Nadu and University Grants Commission, New Delhi.

At present, two mega research projects on Development and Utilization of Flyash as a Carrier in chemical and herbal insecticide formulations to a tune of Rs. 3.21 crores funded by Ministry of Coal and Ministry of Science and Technology, Government of India are underway.

He has to his credit publications of more than 120 research papers and 11 books and organized three national level seminars on 'Use of Flyash in Agriculture a Refresher Course on "Mycoinsecticide : A Potential bio pesticide of Today and Tommarow" and a Refresher Course on 'Organic Pest Control' during 2008. His fields of specialization are:

- Developed a fungal Insecticide, Mycoinsecticide 70% WP against rice brown planthopper and green leafhopper
- Discovered flyash as an insecticide and a carrier in chemical and herbal insecticide formulations
- Documented more than 1000 indigenous pest control practices and revalidated important practices for farmers' use. Also large number of tribal pest control practices collated from various hillocks in Tamil Nadu were also documented and published.